国民築城必携

Building Fortifications : The Citizen's Essential Guide

Japanese Army Manual

ISBN: 978-1-950959-47-1

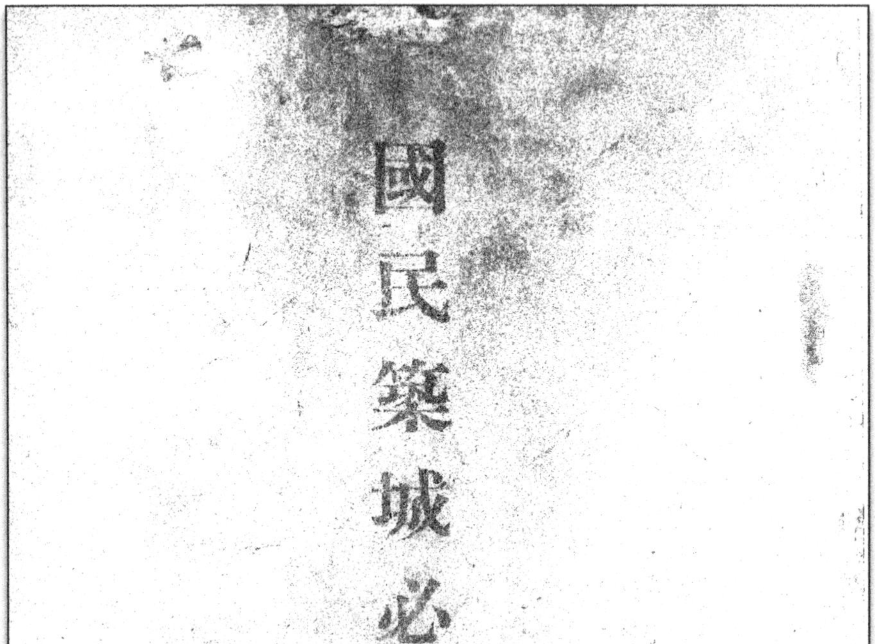

國民築城必携

昭和二十年四月十二日

大本營陸軍部

（增刷許可ス
此ノ場合ハ〇〇複寫トスルコト）

Kokumin Chikujo Hikkei

Building Fortifications : The Citizen's Essential Guide

Published April 12 Showa 20 (1945)
The Imperial General Headquarters Army Division[1]

(Reprinting is authorized. Record the number printed in each run.)

[1] The Imperial General Headquarters was part of the Supreme War Council and was established in 1893 to coordinate efforts between the Imperial Japanese Army and Imperial Japanese Navy during wartime.

要旨

一、築城ハ軍ノ指導ノ下ニ造ルヲ立前トスル本册子ハソノ作リ方ノ參考デアル

二、軍指導以外ノ土地モ皇國守護ノ爲全國民ハ各人ノ郷土ヲ要塞化シ對敵戰鬪ヲ容易ナラシムル爲築城ヲ實施セネバナラヌ

三、築城ノ中最モ大切ナノハ對戰車障碍デアル

四、敵モシ來寇スル時ハ軍ノ指導ニ從ヒ各人ハ其郷土要塞ニ據ツテ戰鬪シ徹底的ニ抗戰セネバナラヌ

Outline

1. While in principle the construction of fortifications will be done under the direction of the military, this guide has been produced as a reference.
2. In order to protect the lands under control of the Emperor, people in areas not under direct military control should fortify their local area. This will make the area easy to defend from enemy assault or launch attacks from.
3. When fortifying an area, the most important thing to build are anti-tank obstacles.
4. In the event an enemy invades, all persons are expected to follow the instructions given by the military and dedicate themselves entirely to the task of defending the fortified local area.

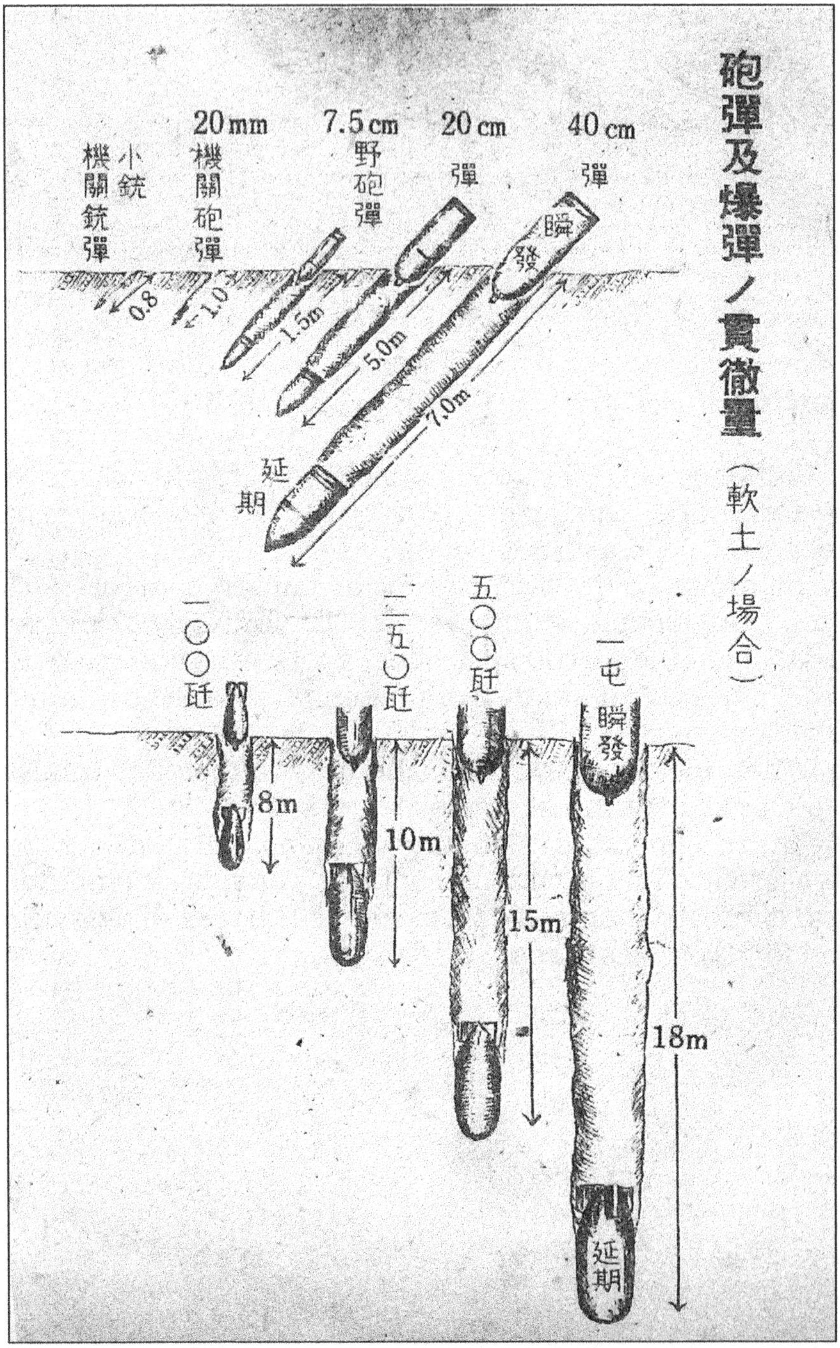
砲彈及爆彈ノ貫徹量
（軟土ノ場合）
小銃機關銃彈
20mm 機關砲彈
7.5cm 野砲彈
20cm 彈
40cm 彈
瞬發
延期
0.8
1.0
1.5m
5.0m
7.0m
一〇〇瓩
二五〇瓩
五〇〇瓩
一屯
8m
10m
15m
18m

Penetrating power of Artillery Shells and Bombs (If on soft earth.)

1. 40 Centimeter Artillery Shell: 7 meters
2. 20 Centimeter Artillery Shell: 5 meters
3. 7.5 Centimeter Field Artillery Shell: 1.5 meters
4. 20 Millimeter Autocannon: 1 meter
5. Small arms or Machine Gun Rounds: 0.8 meters

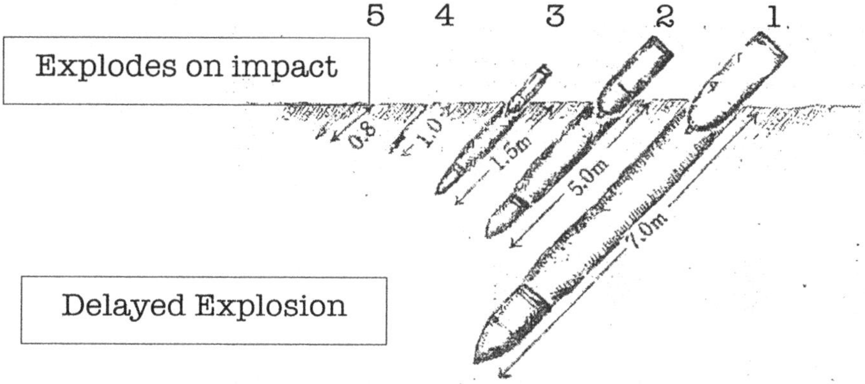

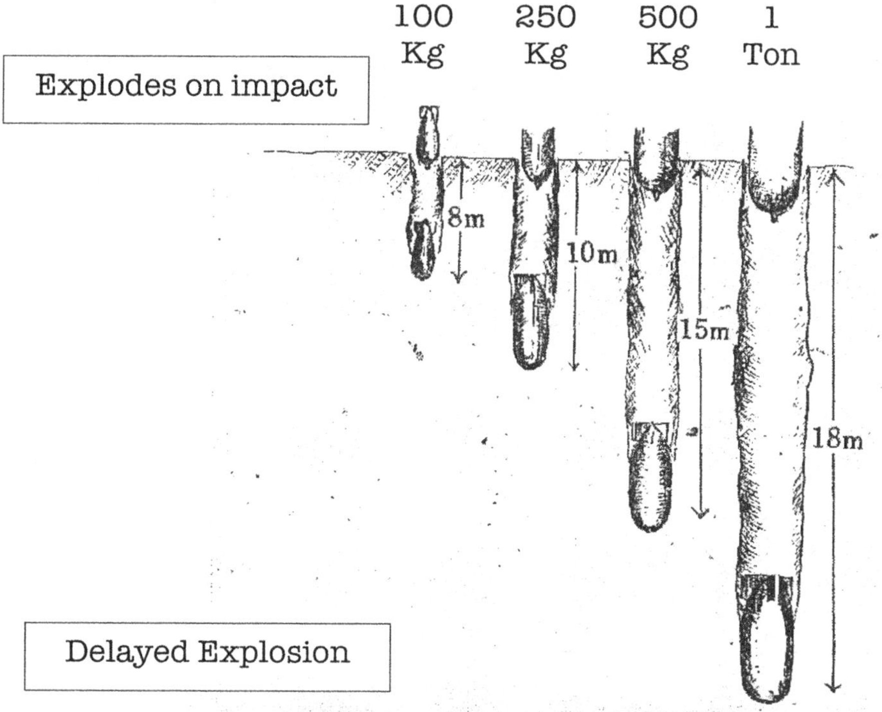

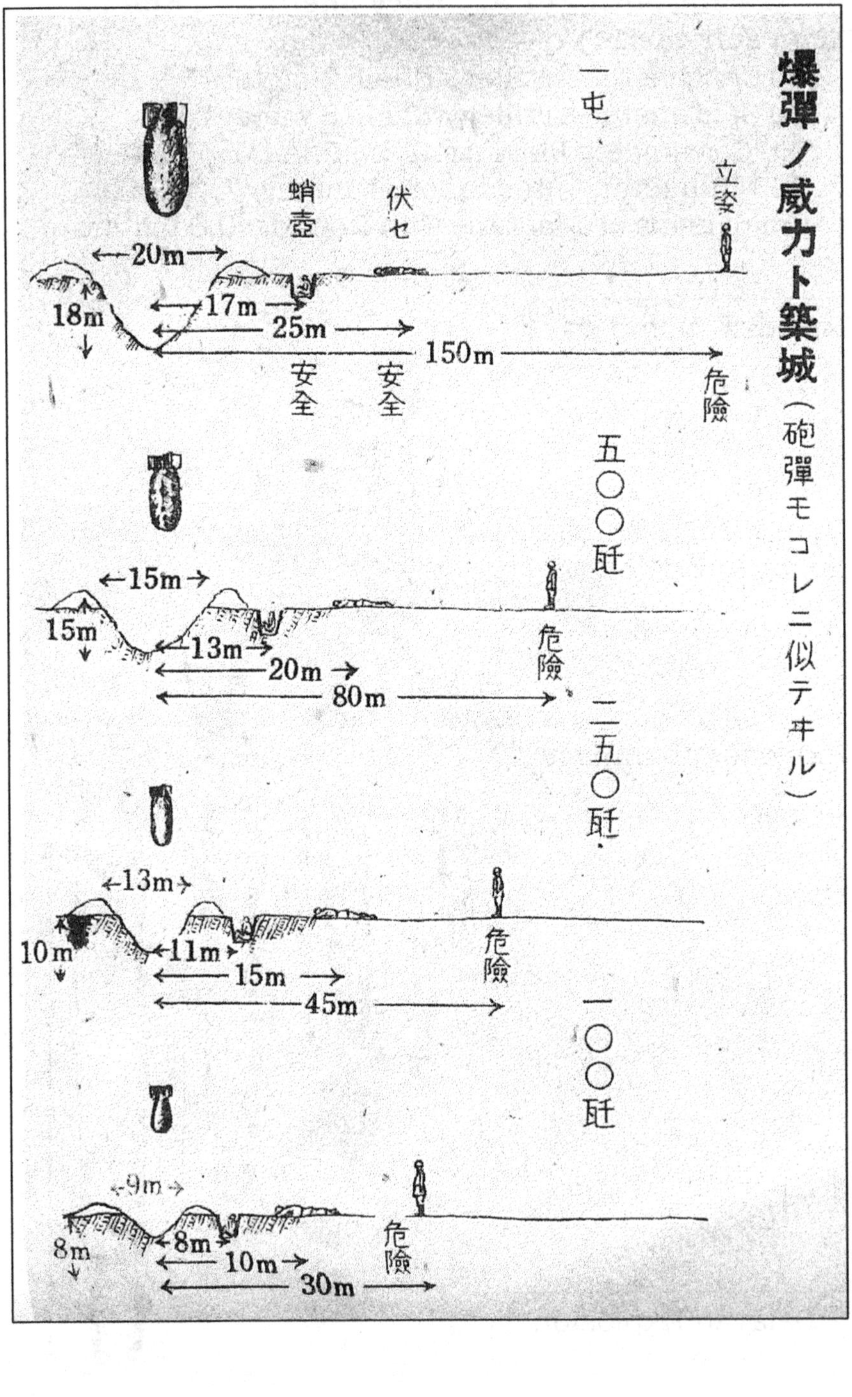
爆弾ノ威力ト築城
（砲弾モコレニ似テヰル）
一屯
蛸壺
伏セ
立姿
20m
18m
17m
25m
150m
安全
安全
危險
五〇〇瓩
15m
15m
13m
20m
80m
危險
二五〇瓩
13m
10m
11m
15m
45m
危險
一〇〇瓩
9m
8m
8m
10m
30m
危險

Bombs Explosive Force and Fortifications
(This also applies to artillery shells)

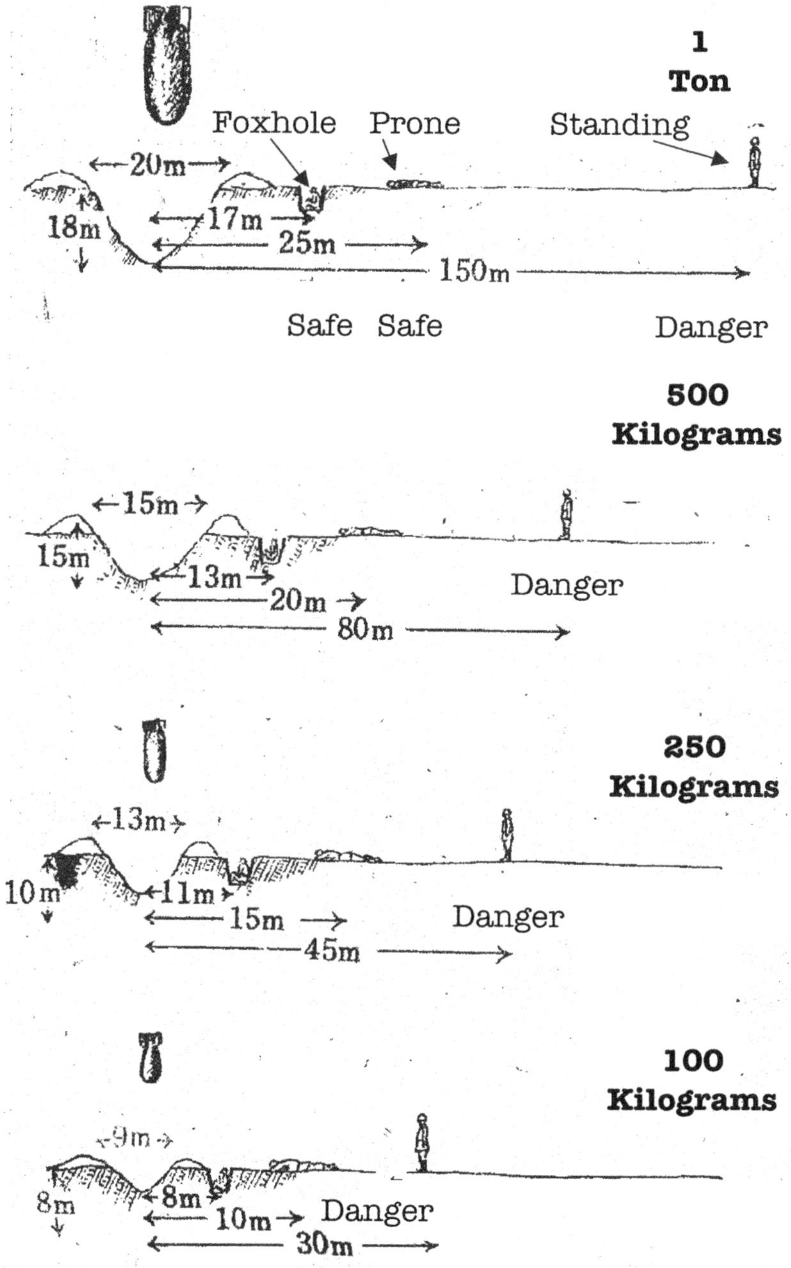

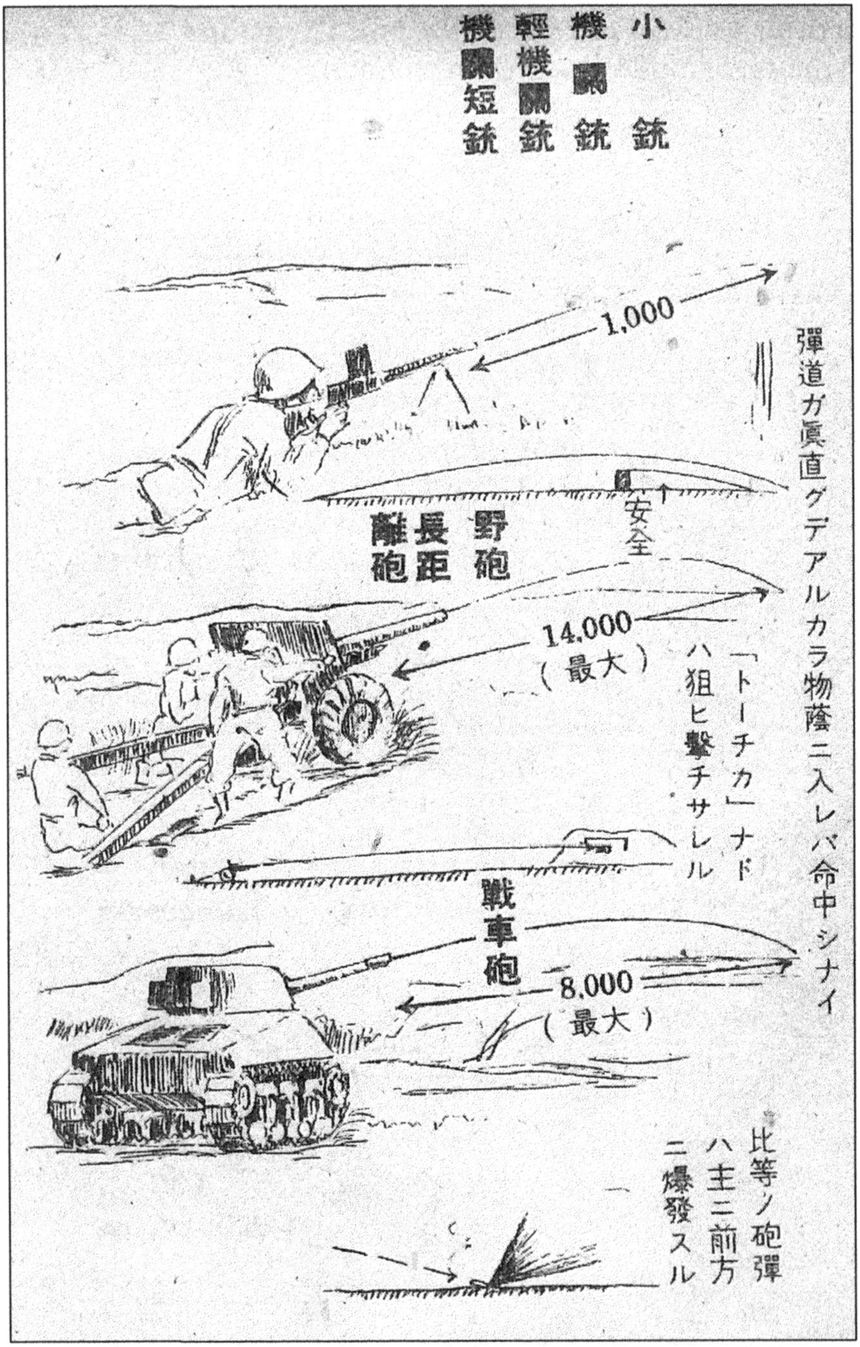
小銃
機關銃
輕機關銃
機關短銃
1,000
安全
野砲
長距離砲
14,000
（最大）
彈道ガ眞直グデアルカラ物蔭ニ入レバ命中シナイ
「トーチカ」ナド
ハ狙ヒ撃チサレル
戰車砲
8,000
（最大）
比等ノ砲彈
ハ主ニ前方
ニ爆發スル

Small Arms · Machine Guns

Light Machine Gun · Sub Machine Gun

These guns have a range of 1000 meters. The bullets travel on a straight path so if you hide behind something, you won't be struck.

Field Artillery · Long Range Gun

(Maximum) Range 14,000 meters

These guns are used to attack Tonchika, or Pillboxes.

*Tonchika is a Russian word used by the Japanese.

Tank Shells

(Maximum Range) 8,000 meters

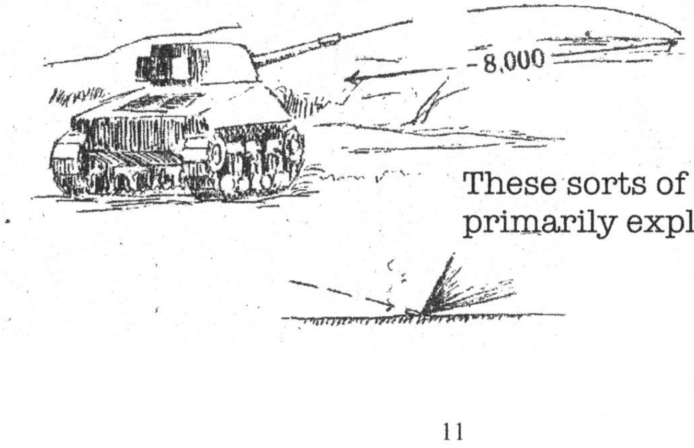

These sorts of shells will primarily explode forward.

ロケット砲
弾道ガ曲ツテキルカラ
山ノ後マデ射テル
焔ノ尾ヲ引クノデ大キ
ク見エル
3000～7000
迫撃砲
此等ノ砲弾ハ
四方ニ爆發スル
（最大）
5000～8000

Rocket Shells

3,000 ~ 7,000 meters

Since rockets travel on an arcing path, they can strike on the far side of mountains. You can observe a fiery tail behind rockets as they fly, which makes them easy to see.

Mortar Shells

(A maximum of) 5,000 ~ 8,000 meters

These types of rounds explode in all directions

艦砲射撃
非常ニ遠クトドク弾丸ノ威力ガ大キイ、シカシ築城ヲシテヰレバ恐ロシクナイ
40cm
砲
30,000m
20 cm
12.7 cm
15,000

Bombardment by Naval Artillery

40-centimeter shells have a range of 30,000 meters
12.7 ~ 20-centimeter shells have a range of 15,000 meters.

These guns can strike targets from an extreme distance and the shells fired have a great destructive power. However, if you have constructed fortifications, there is nothing to fear.

水平爆擊
一齊投下ト連續投下トアル
2,000以上10,000前後

Level Flight Bombing

The bombs can be dropped all at once or in sequence. Generally dropped from above 2,000 meters and below 10,000 meters.

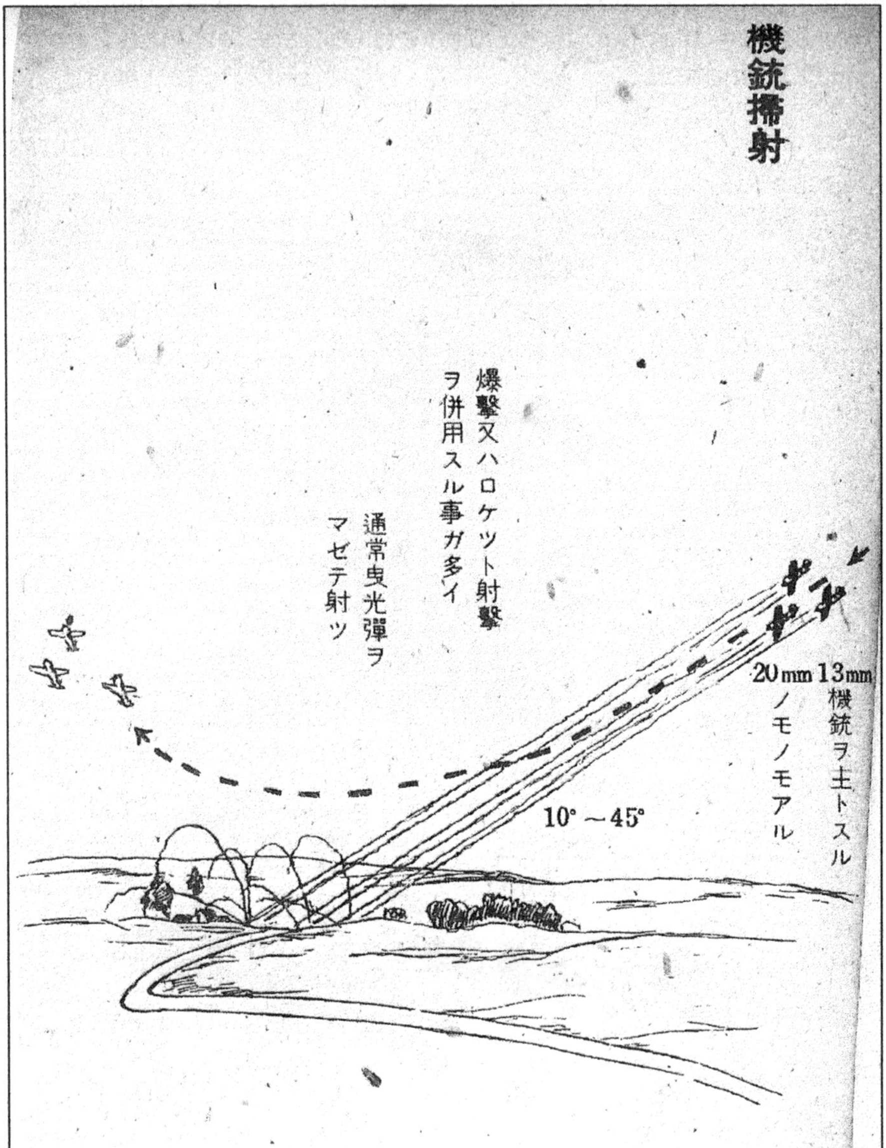
機銃掃射
20mm 13mm
機銃ヲ主トスル
ノモノモアル
爆撃又ハロケット射撃
ヲ併用スル事ガ多イ
通常曳光弾ヲ
マゼテ射ツ
10°～45°

Machine Gun Strafe

- Angle of attack is 10 ~ 45°
- The primary rounds fired are 33-millimeter, however 20-millimeter are also used.
- A machine gun strafe is often done in conjunction with bombing or rocket attacks.
- Typically, tracer rounds are mixed in with the regular ammunition.

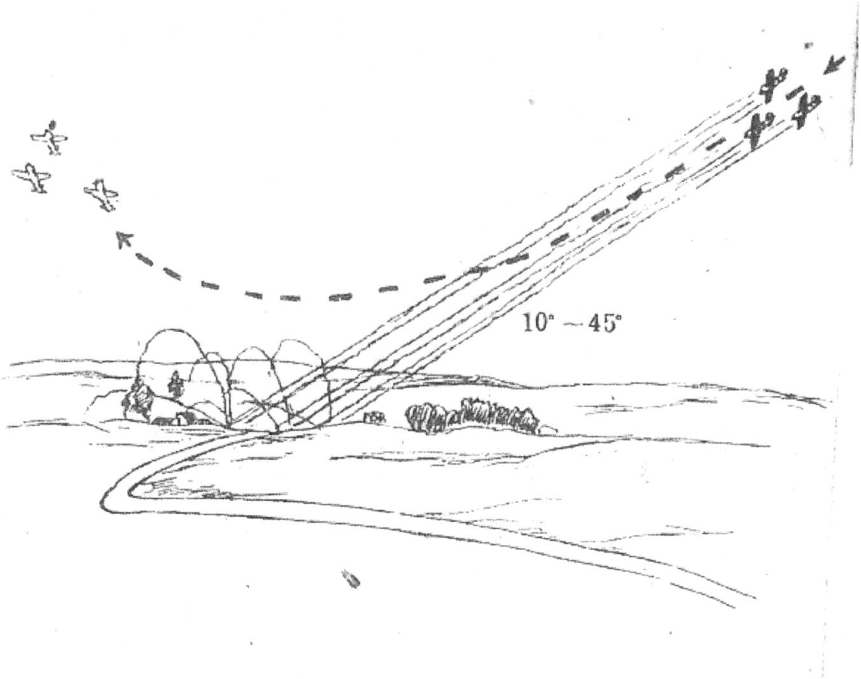

陣地ヲ作ル所
一、部落
戰車ニ隨伴スル
歩兵ヲ射ツ
戰車ハ肉薄攻擊
デ破壞スル

How to Position Your Unit

1. Near a Village

Shoot the troops accompanying the tanks.[2] To destroy the tanks used Nikuhaku Kogeki,[3] close attack.

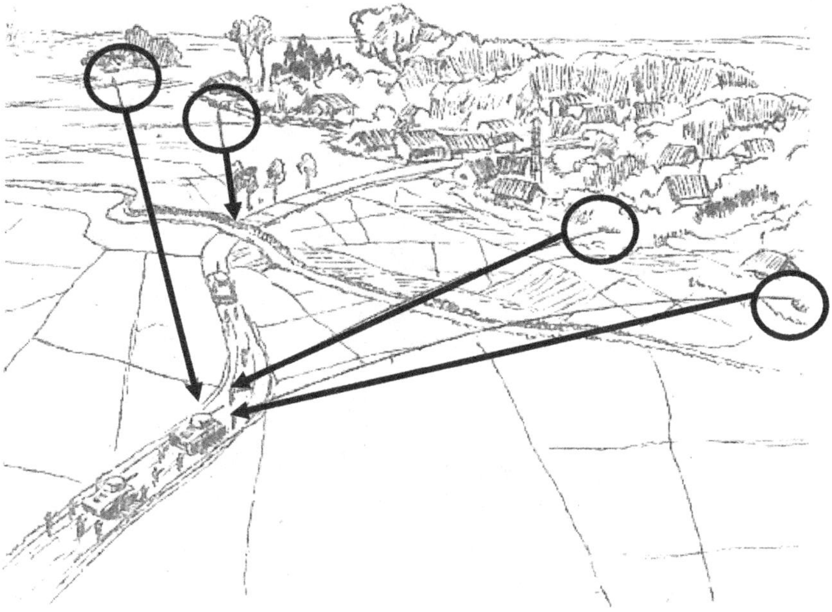

[2] Circles show what seem to be the recommended firing points or sniper positions, though no additional information is given.

[3] Nikuhaku Kogeki is comprised of two Kanji, Meat 肉 and Thin 薄. It means to be close enough so that your flesh is pressing against the enemy. In this case a tank.

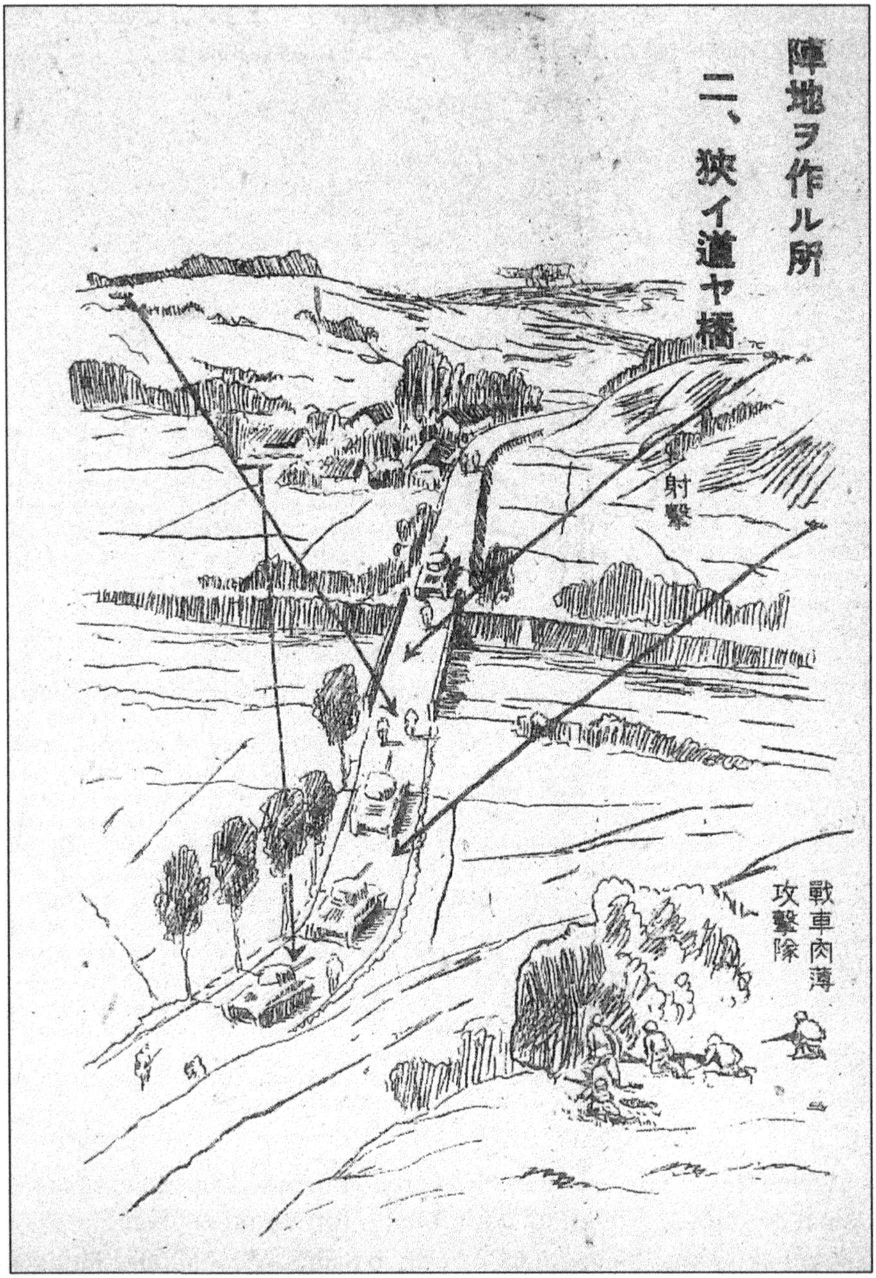
陣地ヲ作ル所
二、狭イ道ヤ橋
射撃
戰車肉薄
攻擊隊

How to Position Your Unit

#2 At a Narrow Road or Bridge

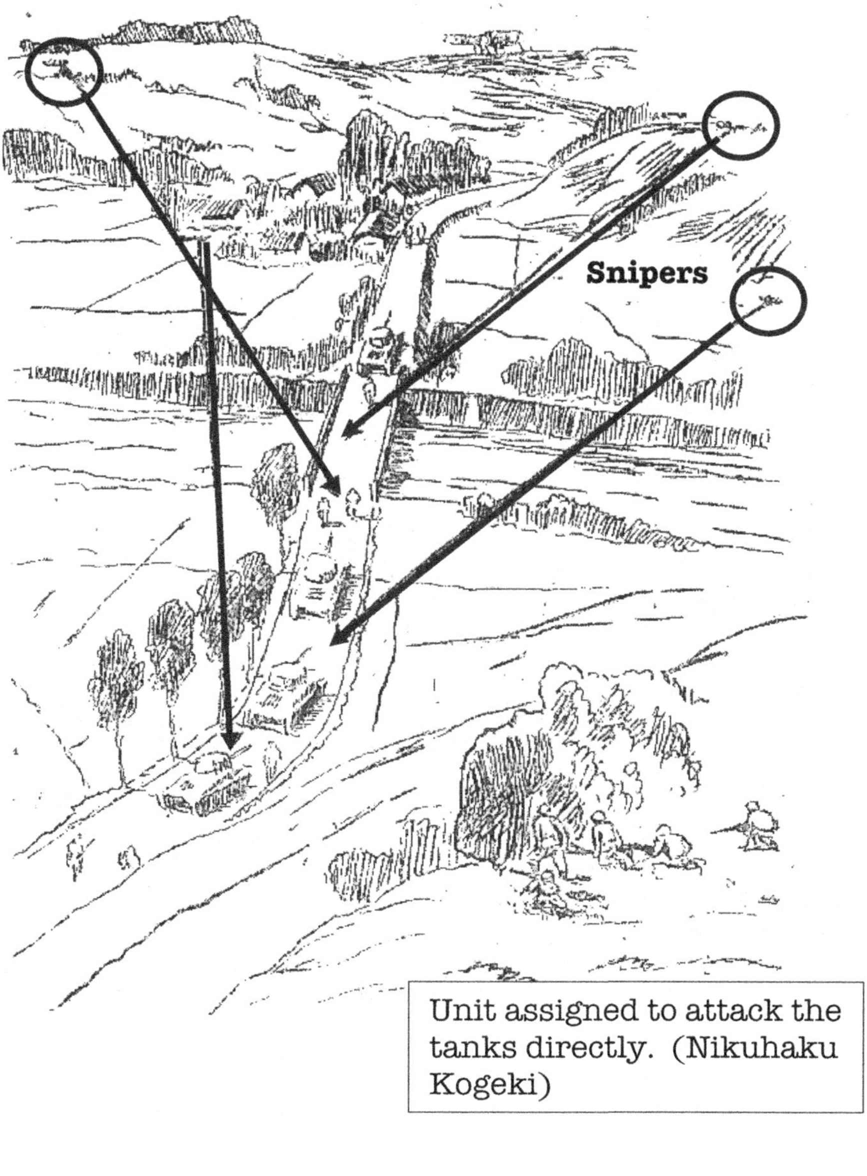

陣地ヲ作ル所
三、盆地
敵ノ前進ヲ待チ構ヘテ四方カラ一齊ニ攻撃スル

How to Position Your Unit #3 Around a Valley

Wait for the enemy to advance on your position and then attack simultaneously from four directions.

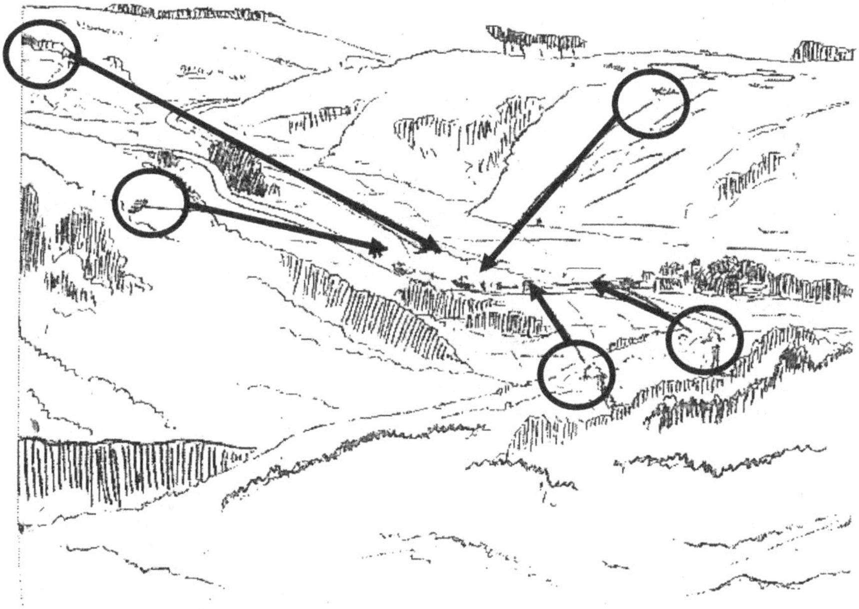

障地ヲ作ル所
四、山道、森林
道ノ曲リ角ニ障碍物（石、防材等）
ヲ設ケ又道路ヲ
破壊スル

How to Position Your Unit
#4 On a Mountain Road or in the Forest

Place obstacles at the turns in the road. (Use rocks, lumber or other materials) You can also destroy the road at that spot.

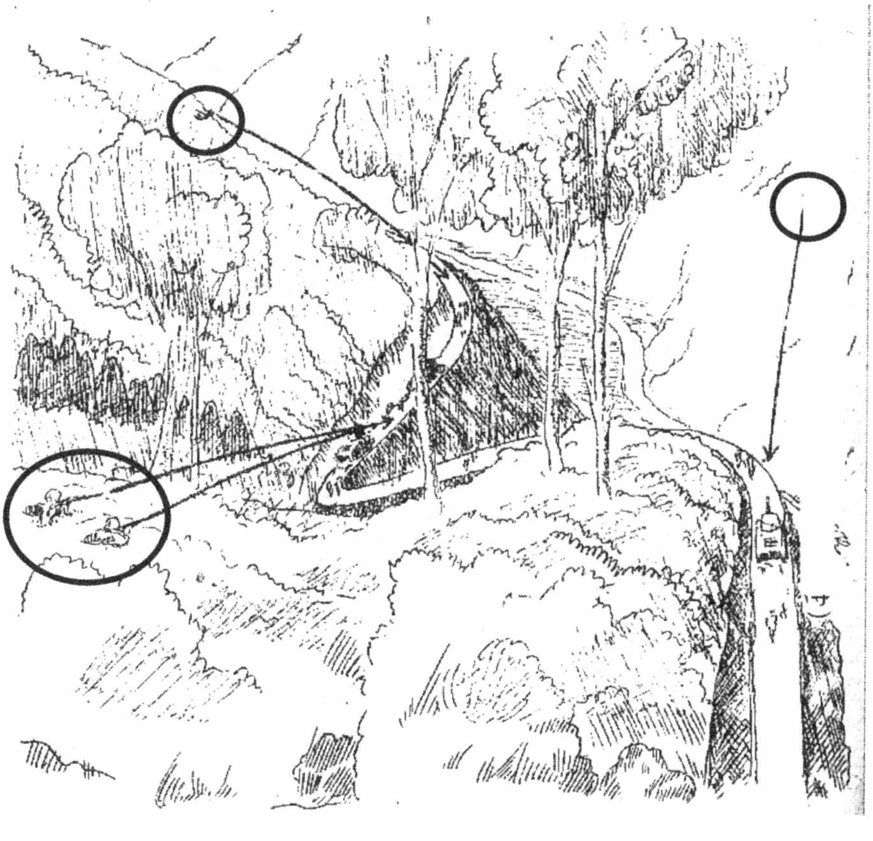

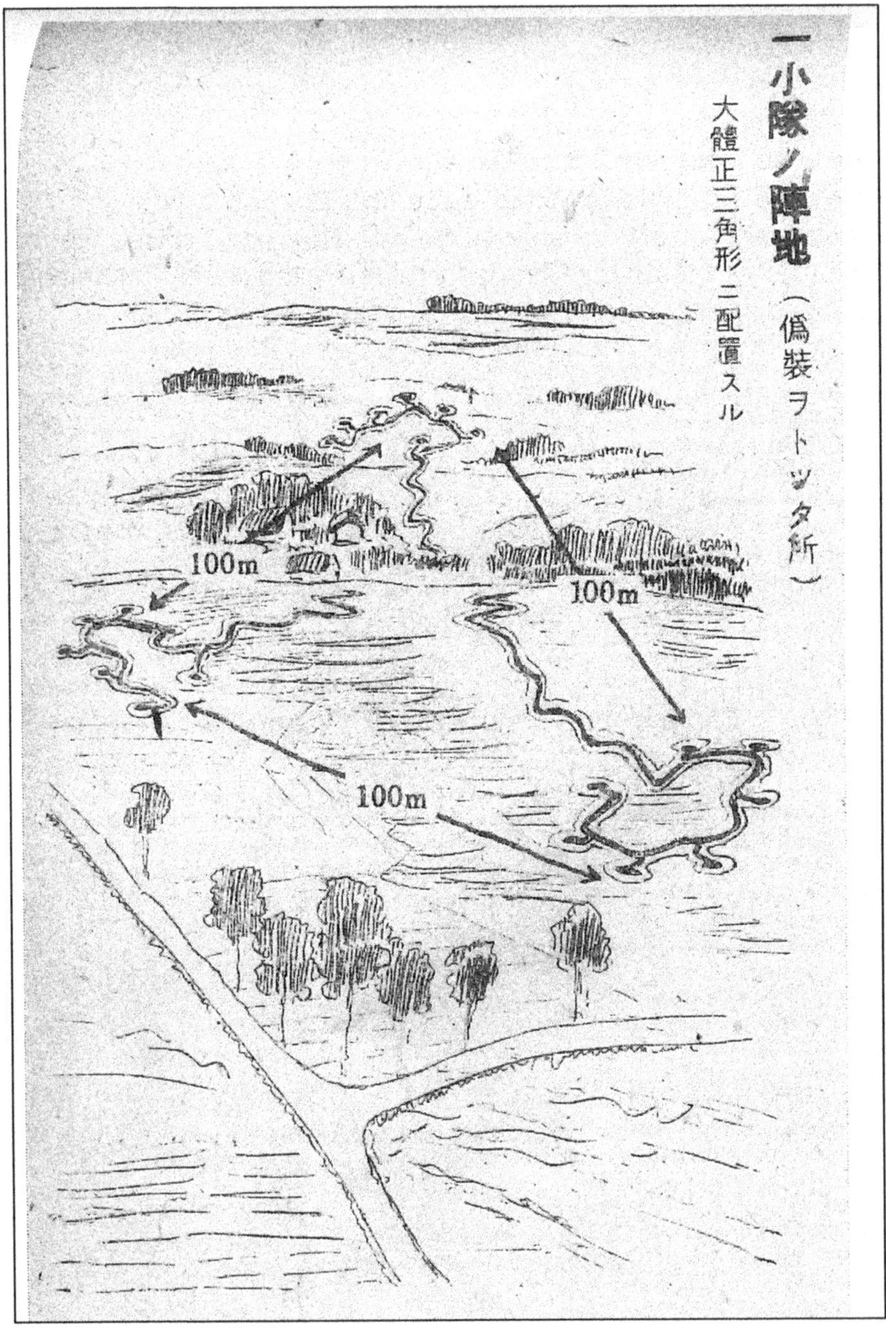
一小隊ノ陣地（偽装ヲトツタ所）
大體正三角形ニ配置スル
100m
100m
100m

How to Position One Small Unit
(A Camouflaged Position)
Your position should be laid out in a triangle.

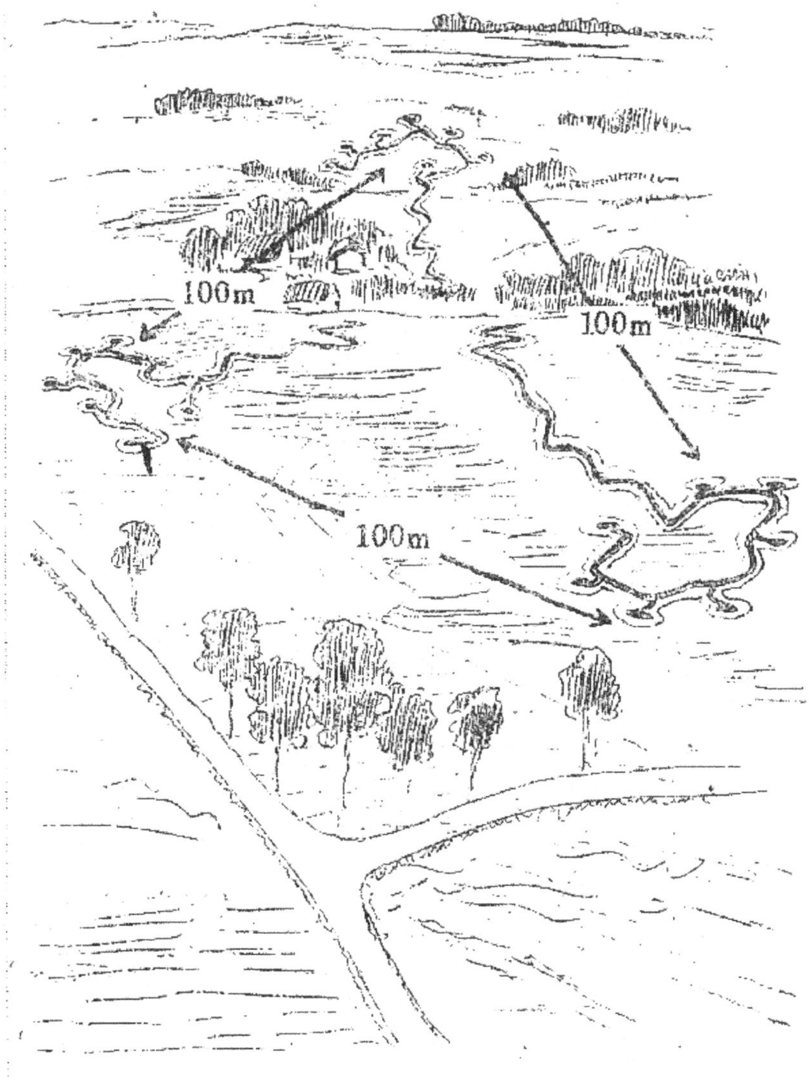

一分隊ノ陣地（偽装ヲトツタ所）
6m

How One Unit Should Be Distributed

(A Camouflaged Position)

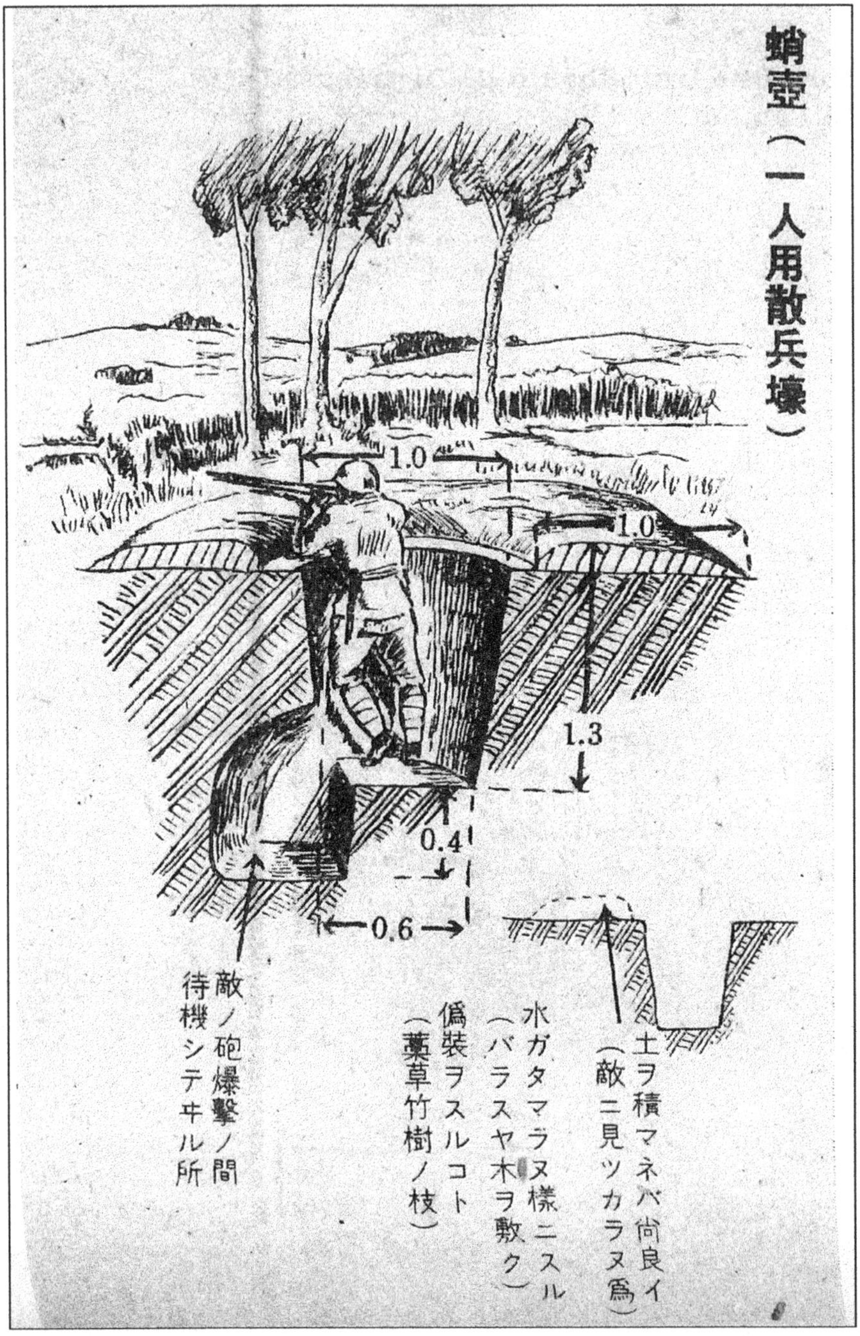

蛸壺（一人用散兵壕）
1.0
1.0
1.3
0.4
0.6
土ヲ積マネバ尚良イ
（敵ニ見ツカラヌ爲）
水ガタマラヌ樣ニスル
（バラスヤ木ヲ敷ク）
僞裝ヲスルコト
（藁草竹樹ノ枝）
敵ノ砲爆撃ノ間
待機シテヰル所

Tako Tsubo[4] Foxhole

A One-Person Soldier's Air Raid Shelter

Be sure to prevent water from accumulating.
(Spread gravel or wood on the ground.)
Use camouflage.
(Rice straw, grass, bamboo or tree branches.)

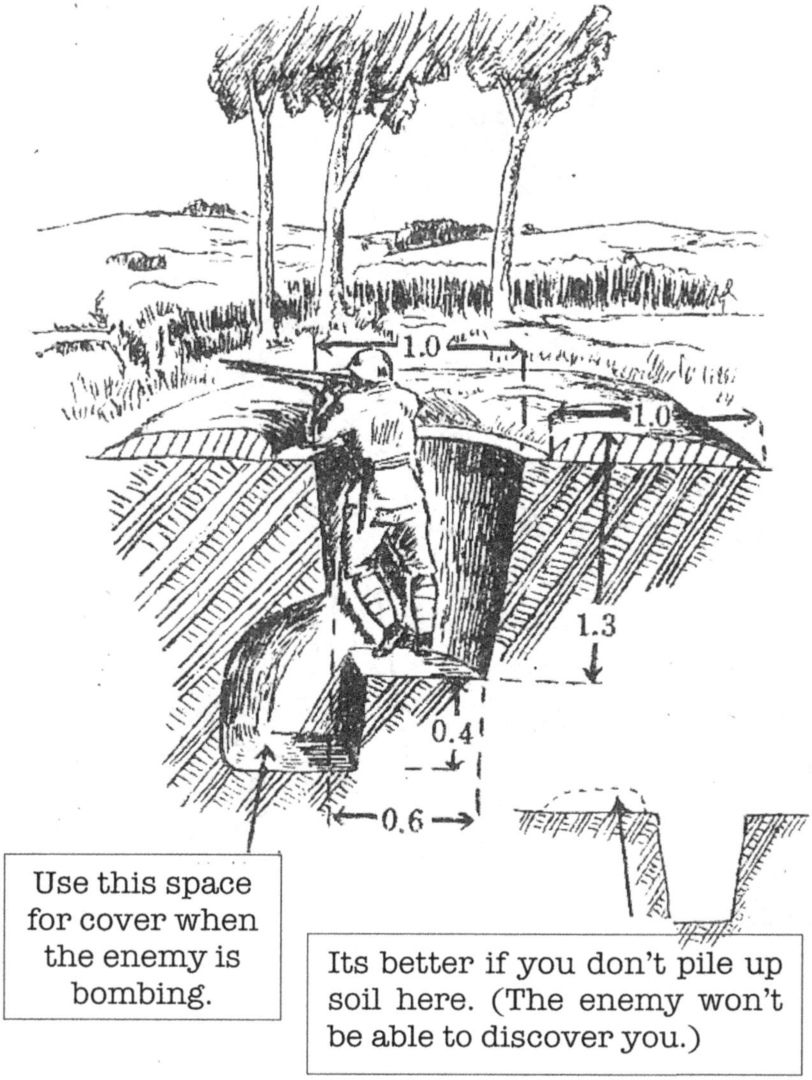

[4] Tako Tsubo is an "octopus pot" a tall earthenware pot submerged in the ocean to catch octopus.

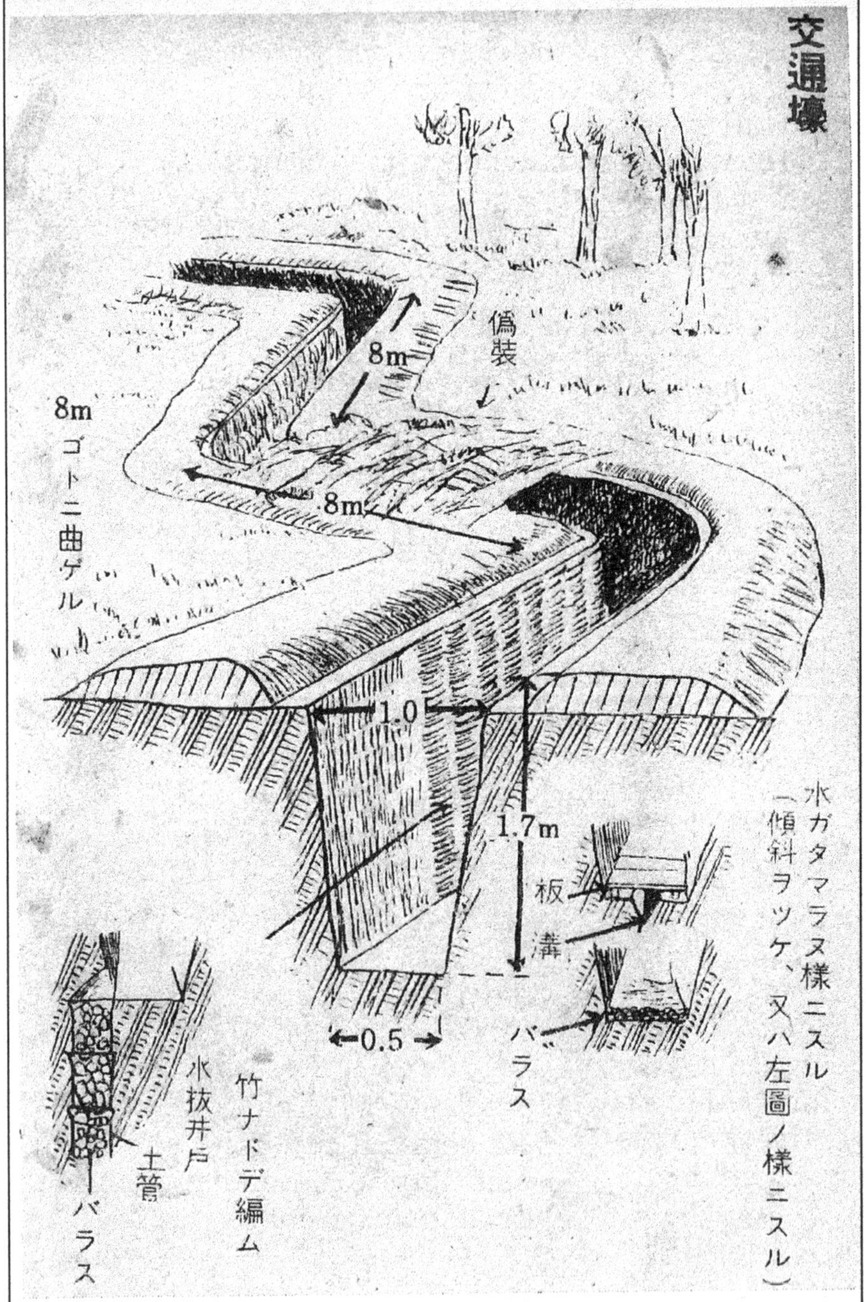
交通壕
8m
偽装
8m
ゴトニ曲ゲル
8m
1.0
1.7m
板
溝
0.5
バラス
水ガタマラヌ様ニスル
（傾斜ヲツケ、又ハ左圖ノ様ニスル）
水抜井戸
竹ナドデ編ム
土管
バラス

Interconnecting Trenches

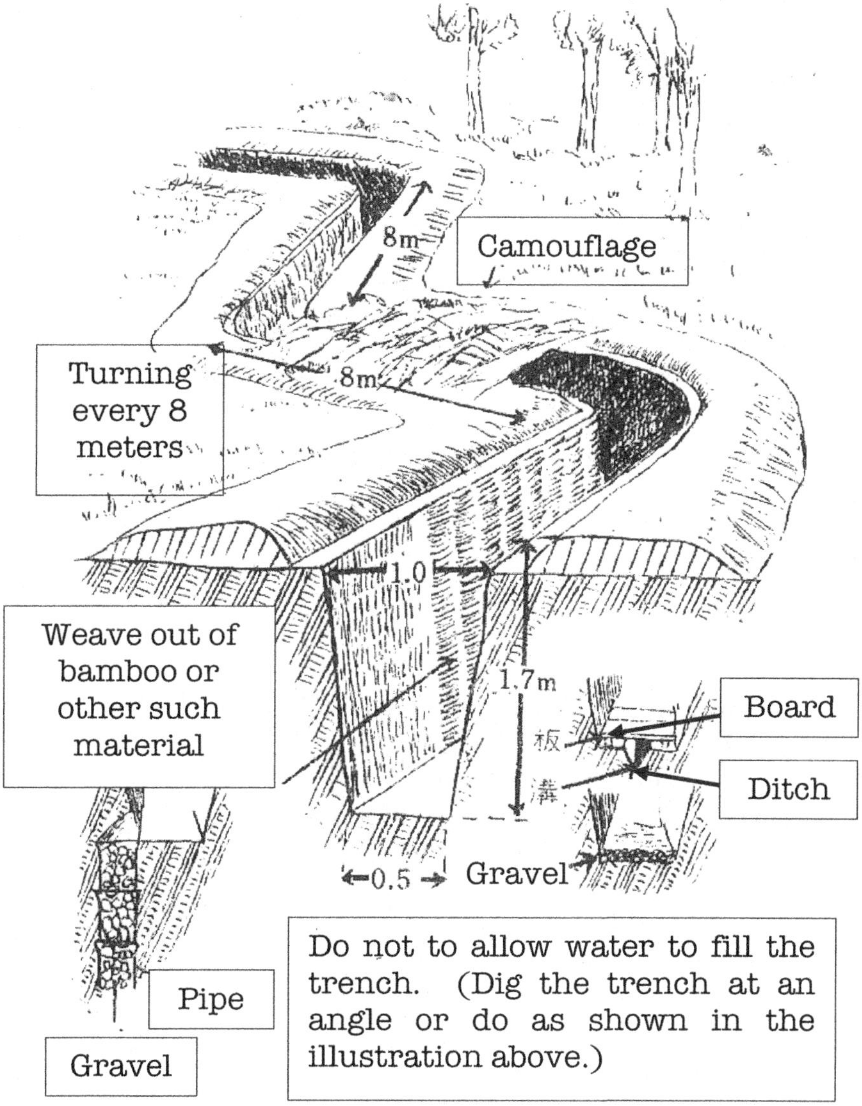

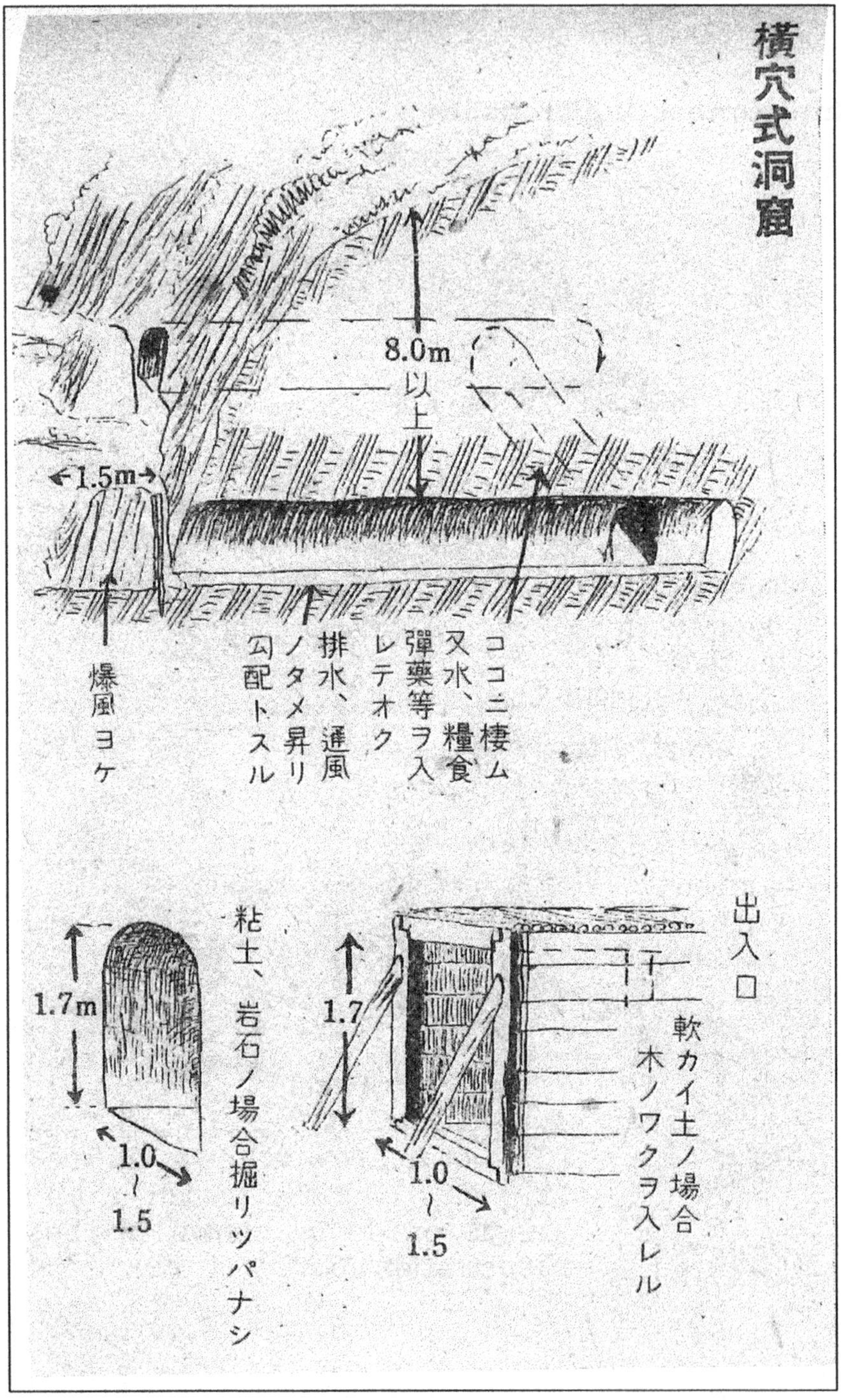
横穴式洞窟
8.0m以上
1.5m
ココニ棲ム
又水、糧食
弾薬等ヲ入
レテオク
排水、通風
ノタメ昇リ
勾配トスル
爆風ヨケ
出入口
軟カイ土ノ場合
木ノワクヲ入レル
1.7
1.0〜1.5
粘土、岩石ノ場合掘リッパナシ
1.7m
1.0〜1.5

Horizontal Cave

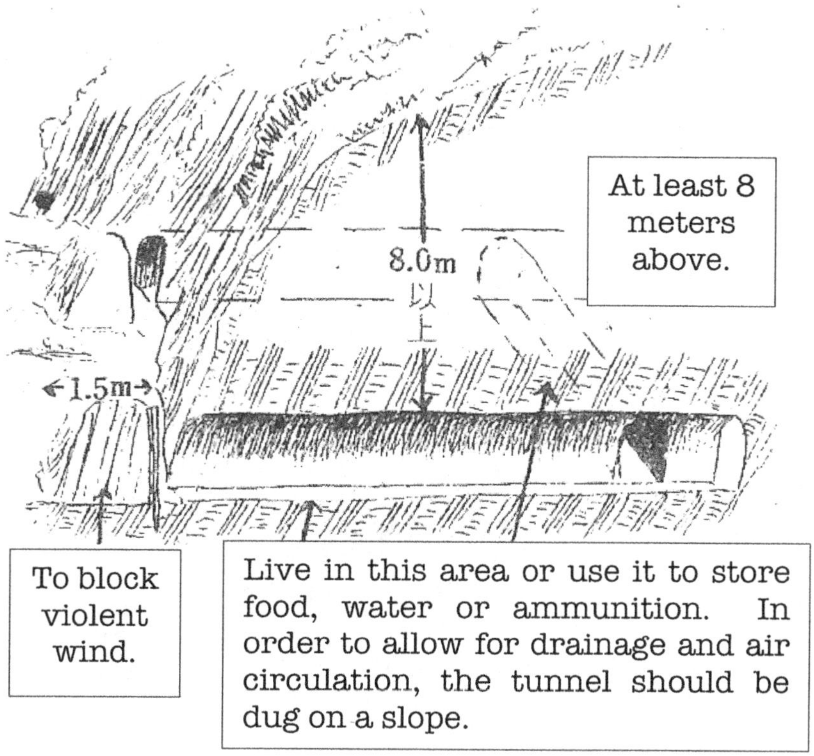

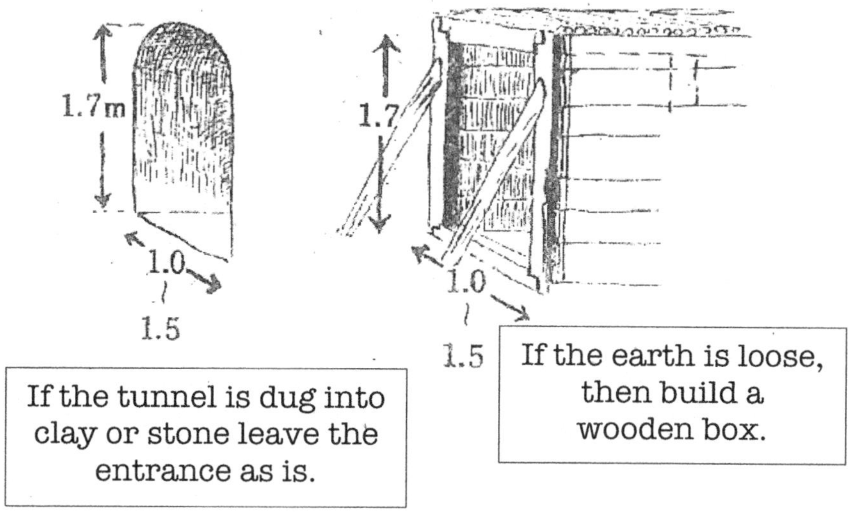

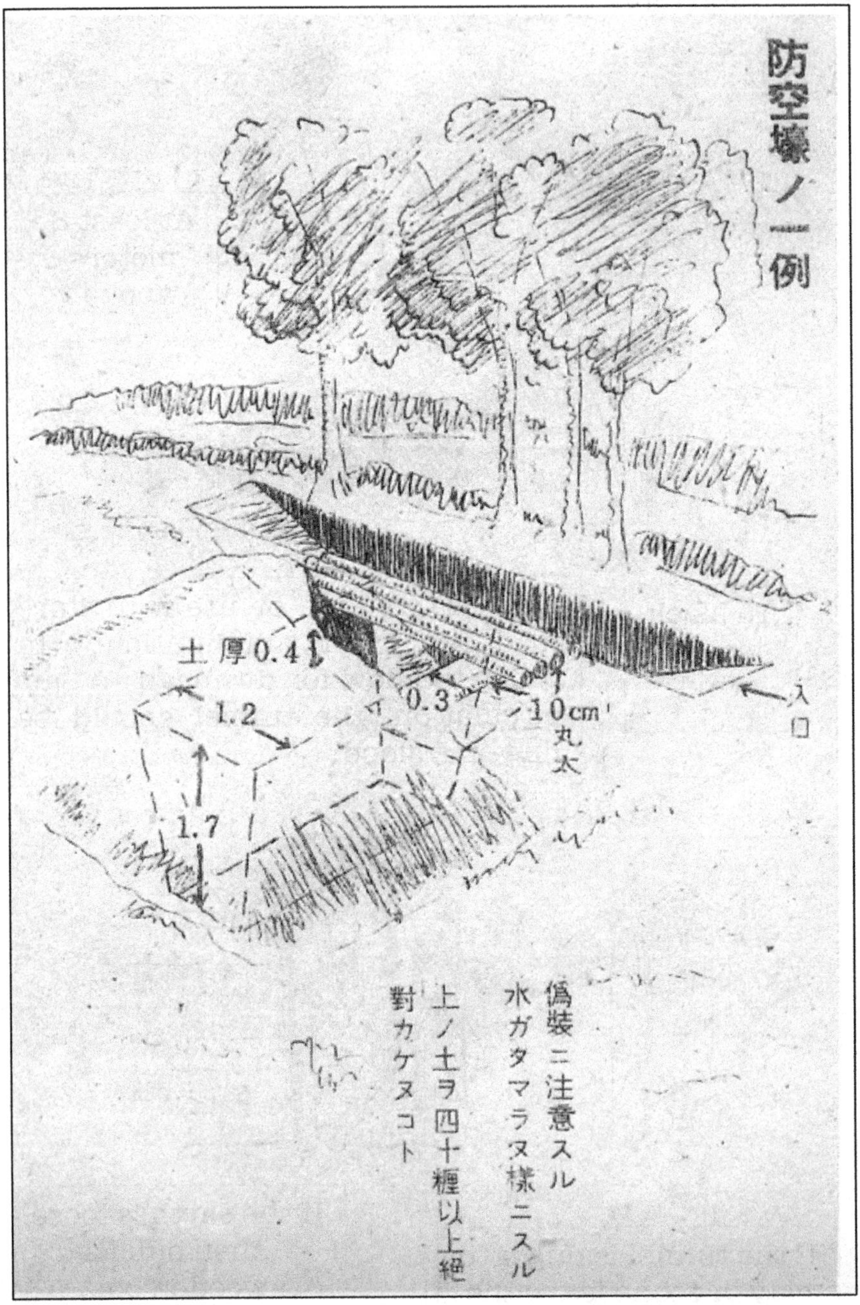
防空壕ノ一例
土厚0.4
1.2
1.7
0.3
10cm
丸太
入口
偽装ニ注意スル
水ガタマラヌ様ニスル
上ノ土ヲ四十糎以上絶
對カケヌコト

An Example of an Air Raid Shelter

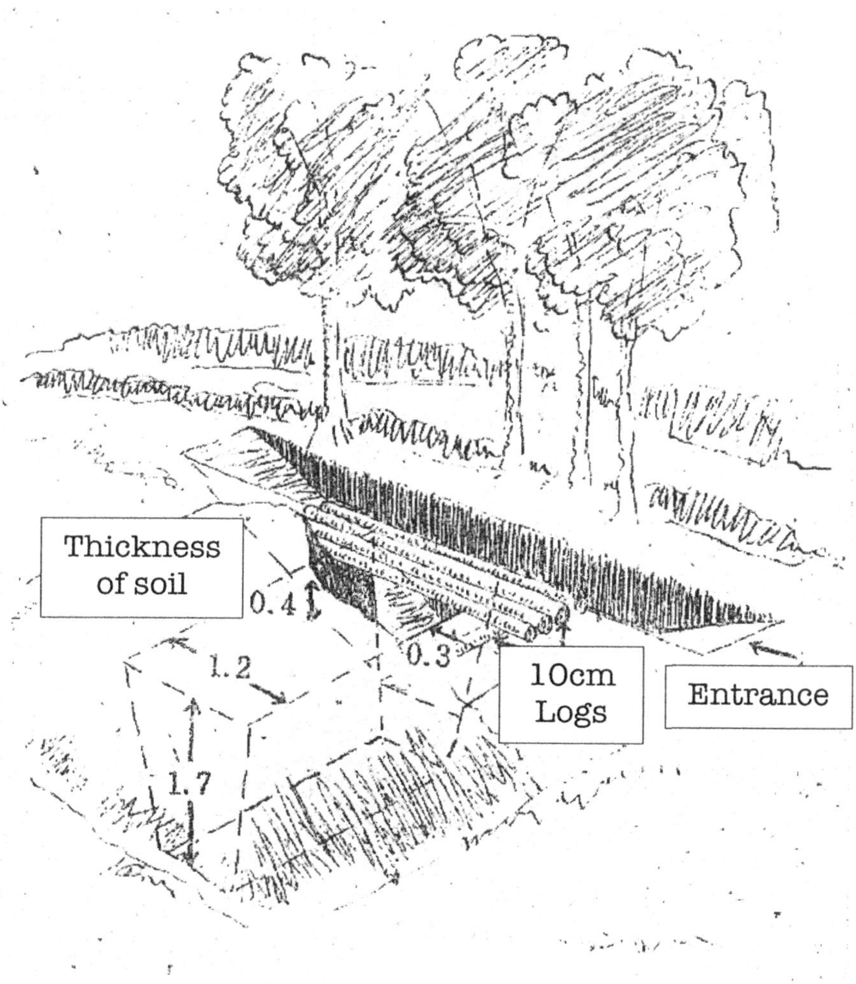

- Be careful how you camouflage the shelter.
- Do not allow water to pool at the bottom.
- Be absolutely certain there is at least 40 centimeters of soil above you.

對戰車障碍物（其一）

1. 道ヤ土手ノ片方ヲ切ル

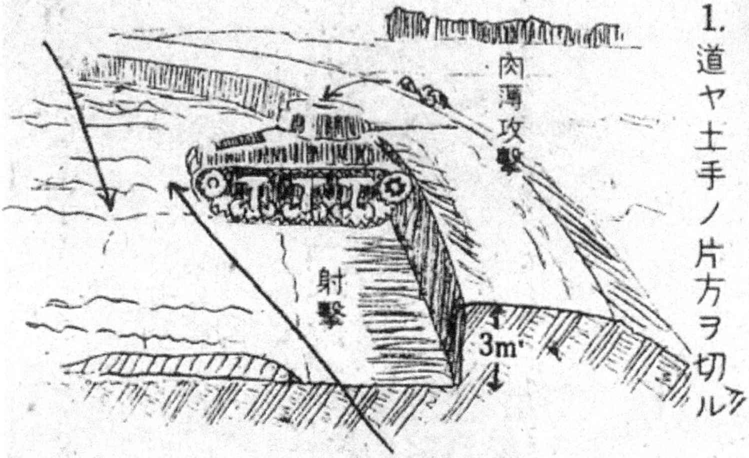

2. 畠地、生垣、林ノ端ニ三角形ノ戰車壕ヲ掘ル

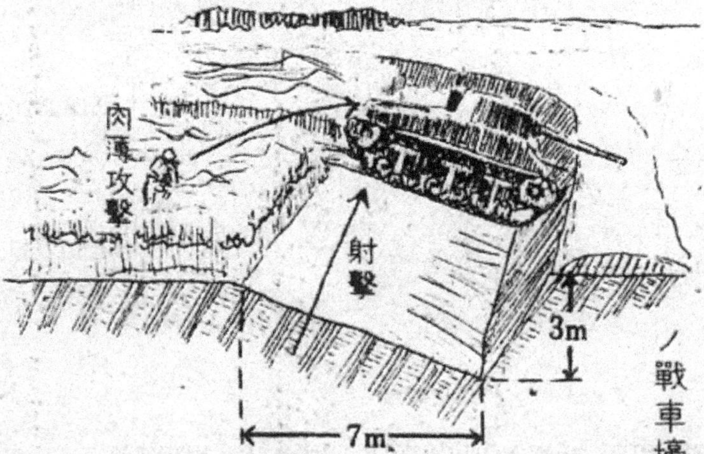

Building Tank Obstacles (Part 1)

1. Dig out a road or embankment on one side.

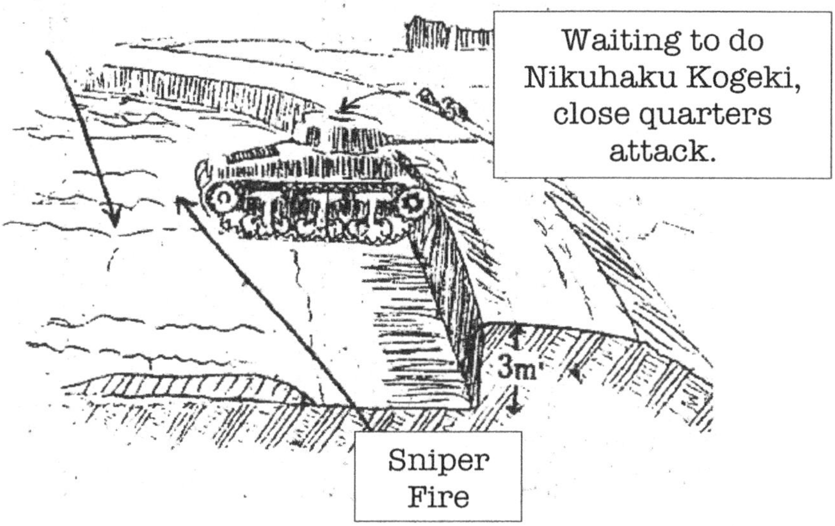

2. At the edge of a farm, section or undergrowth or wooded area, dig a triangular tank bunker.

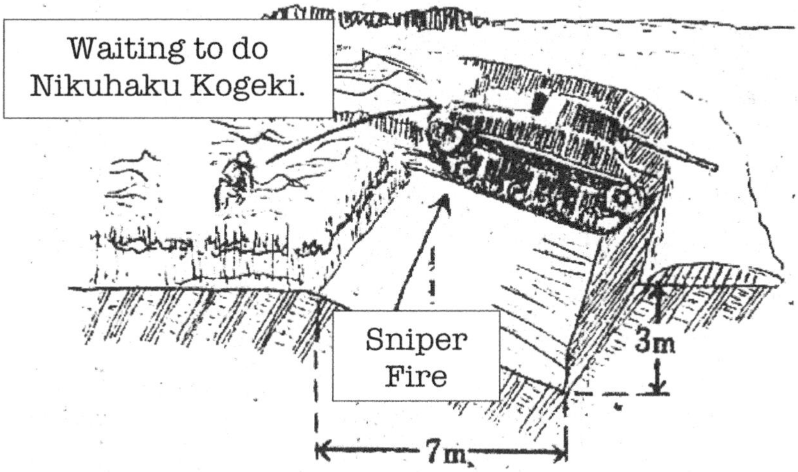

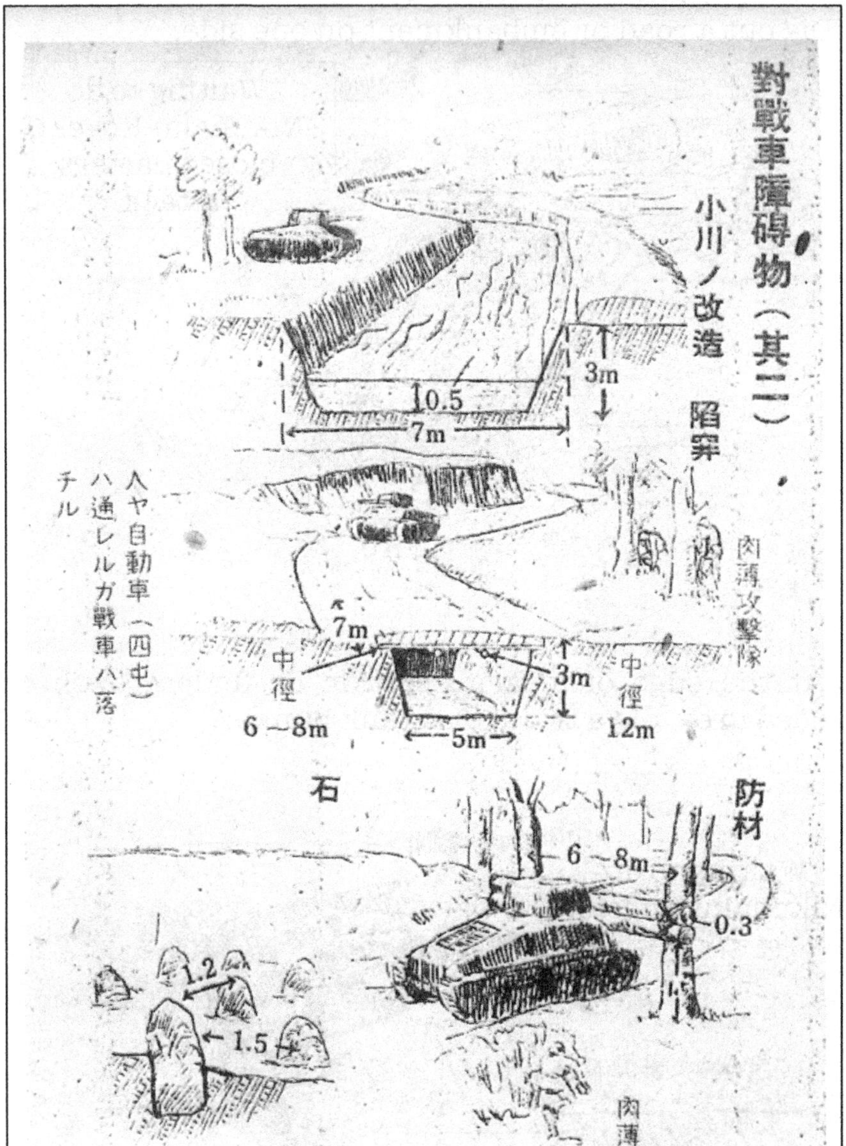
對戰車障碍物（其二）
小川ノ改造
3m
0.5
7m
陷穽
肉薄攻擊隊
人ヤ自動車（四屯）ハ通レルガ戰車ハ落チル
7m
中徑
6～8m
3m
5m
中徑
12m
石
防材
6～8m
0.3
1.2
1.5
肉薄攻擊

Building Tank Obstacles (Part 2)

Redirect a small river.

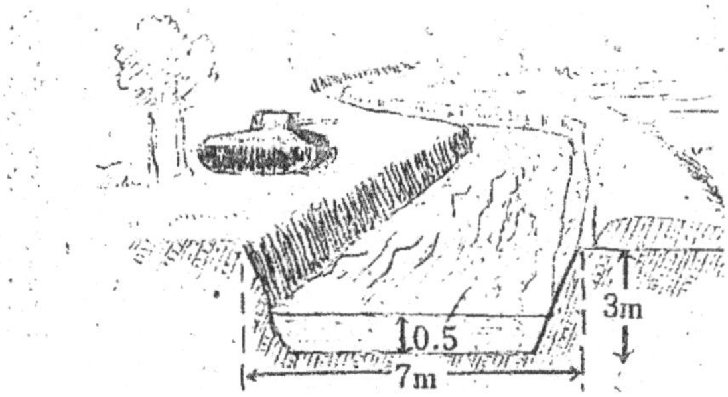

Make an Otoshi Ana, Deadfall

People or vehicles (up to 4 tons) can pass, but tanks will fall.

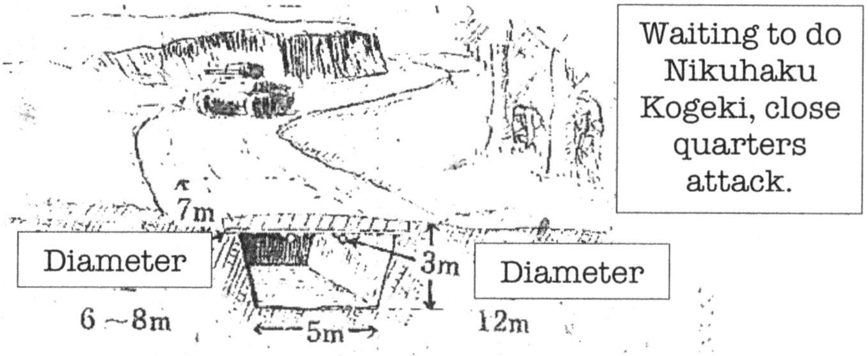

Defensive materials

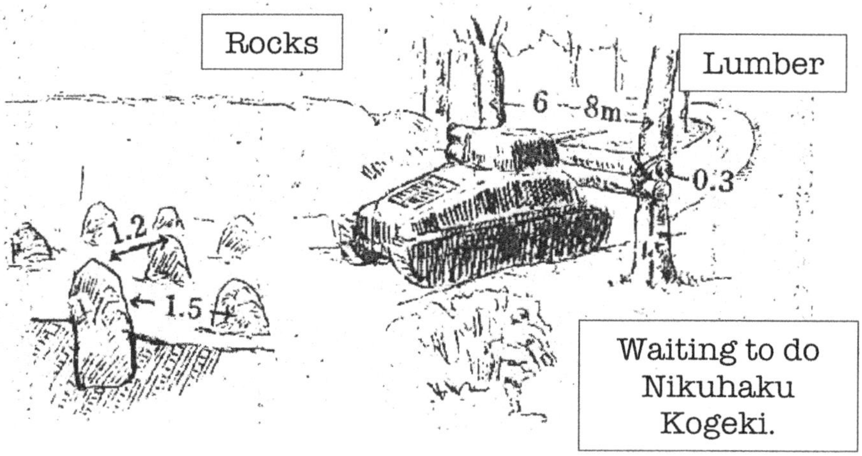

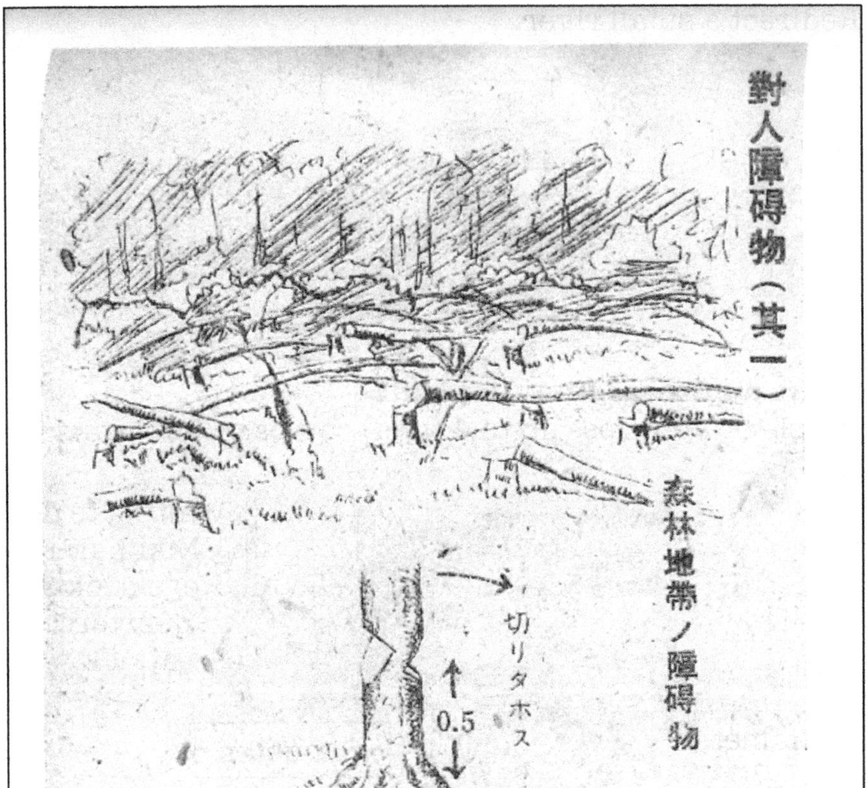
對人障碍物（其一）
森林地帶ノ障碍物
切リタホス
0.5
先ヲトガラス
逆茂木
0.7
0.5
1.0
固定スル

Anti-personnel Obstacles (Part 1)

How to make anti-personnel obstacles in the forest.

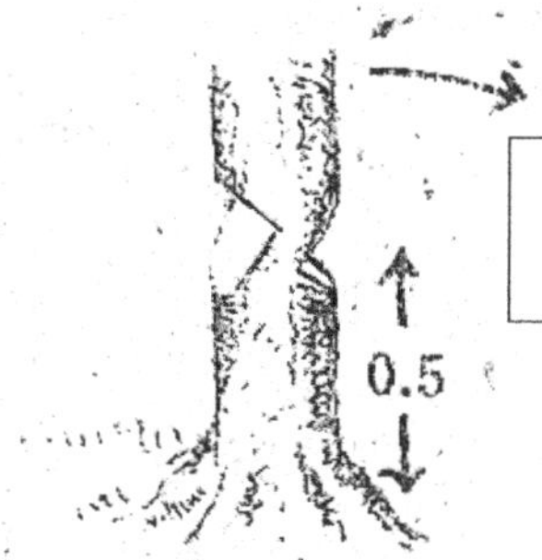

Cut and topple trees as shown.

Break off the ends of branches so they are sharp.

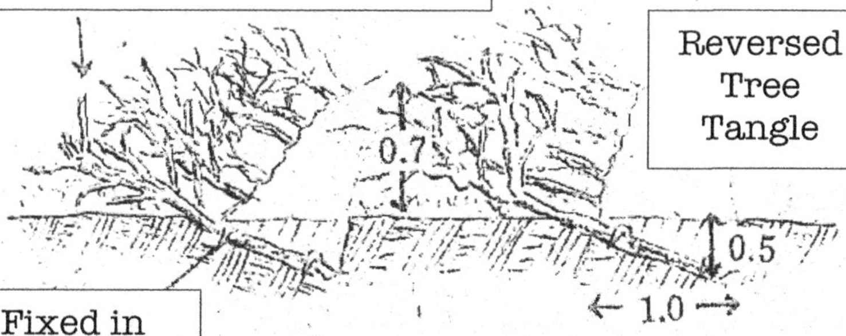

Reversed Tree Tangle

Fixed in the ground.

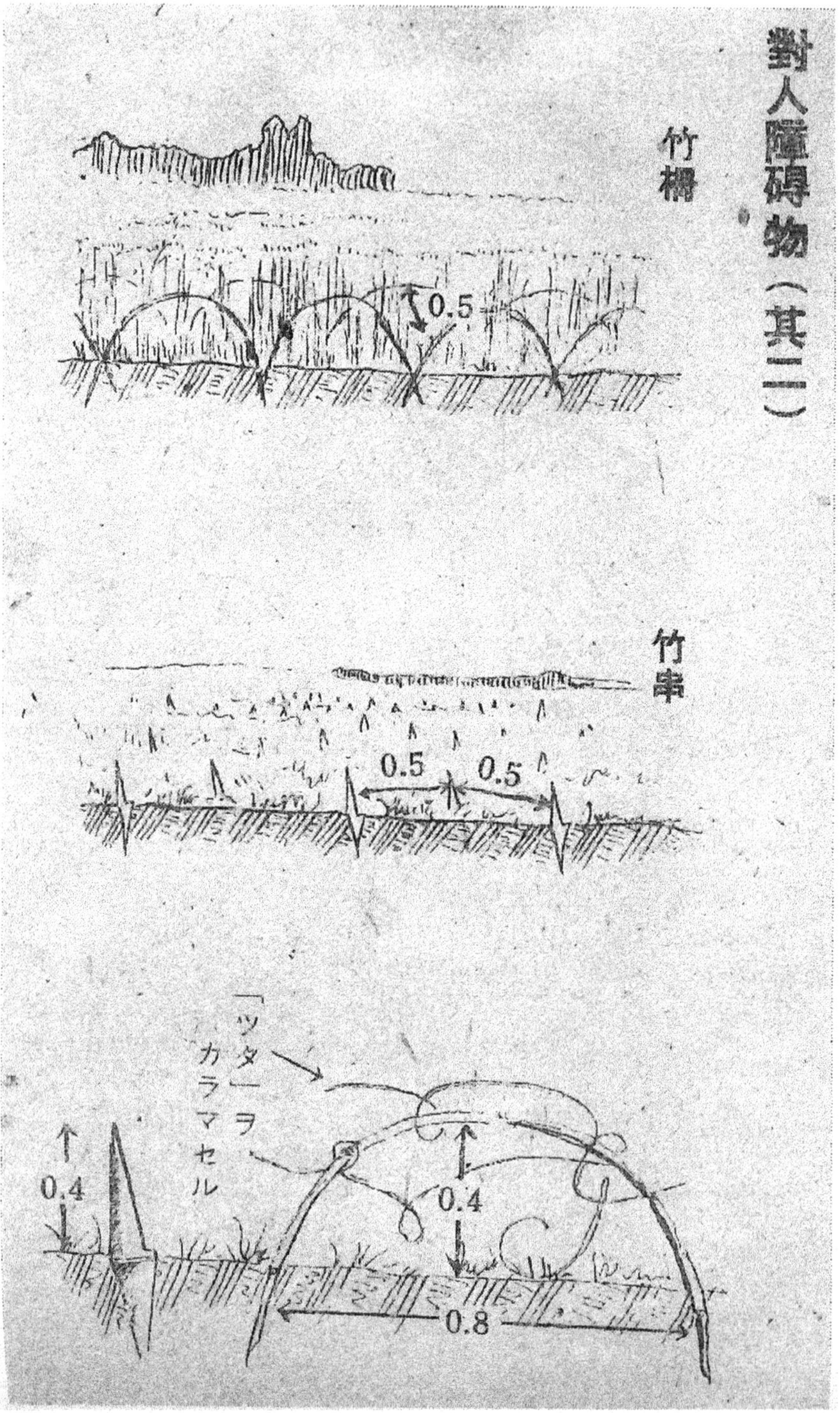
對人障碍物（其二）
竹柵
0.5
竹串
0.5
0.5
「ツタ」ヲカラマセル
0.4
0.4
0.8

Anti-personnel Obstacles (Part 2)

Bamboo Fence

Bamboo Skewers

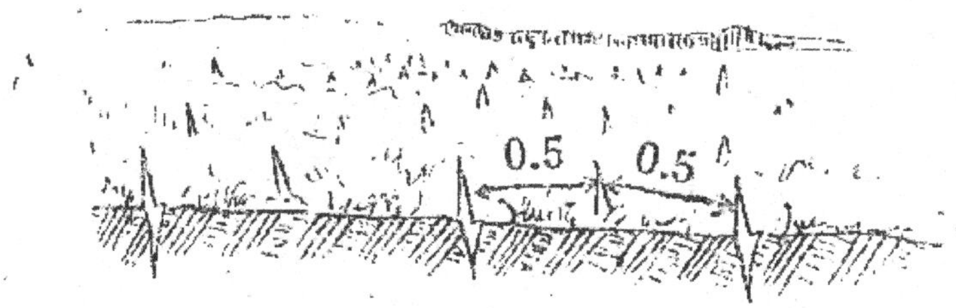

Conceal the fence with ivy.

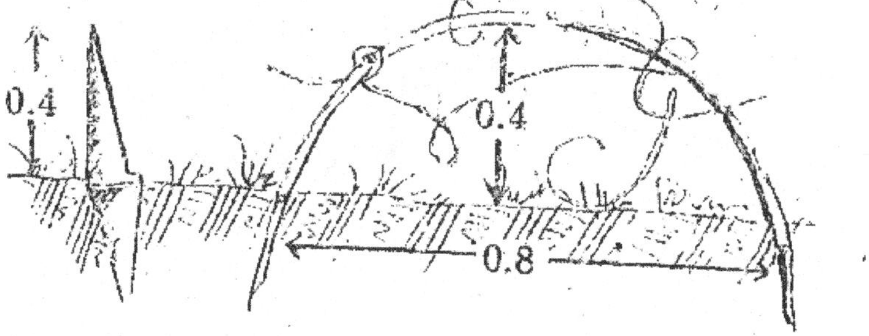

斬込隊ノ潜伏根據地
森林内
糧食、水、弾薬等ヲ入レル
入口ガワカラヌヤウニスル
平地ノ草ムラ
開キフタ
藁ナドノ下
藁堆
馬小屋ナド
納屋

Undercover Base for the Kirikomitai, Cutting In Division/ Surprise Attack Division

In the forest.

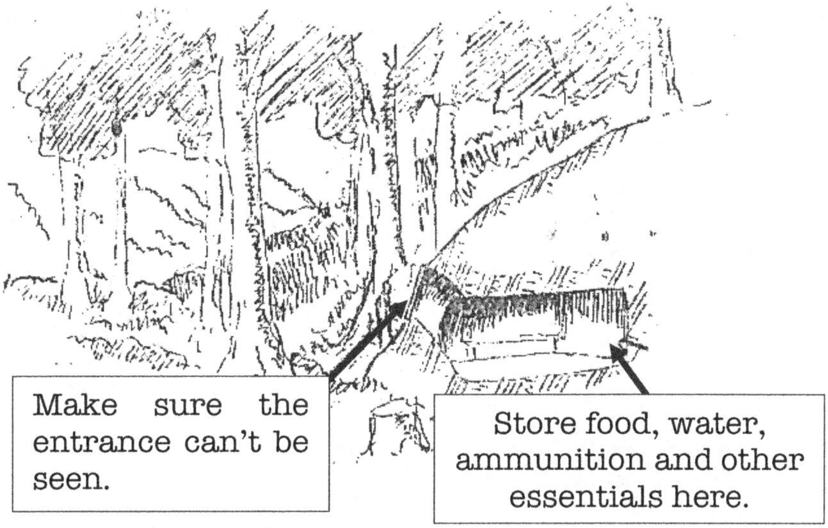

Under a Haystack

In a Flat Grassy Area

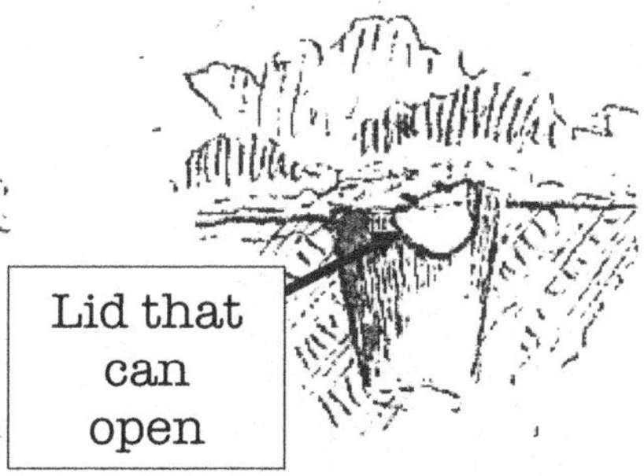

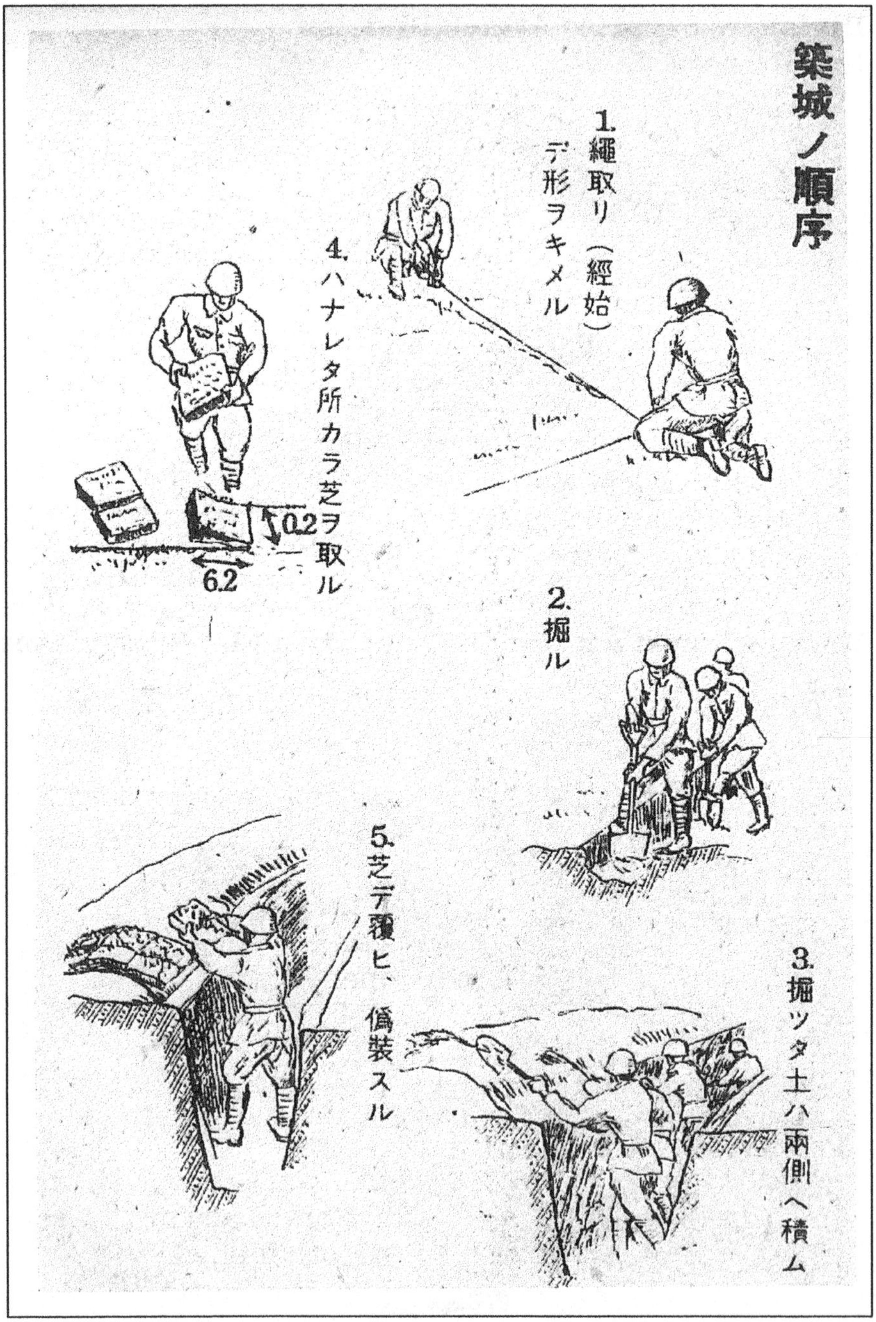
築城ノ順序
1. 繩取リ（經始）デ形ヲキメル
4. ハナレタ所カラ芝ヲ取ル
0.2
6.2
2. 掘ル
5. 芝デ覆ヒ、僞裝スル
3. 掘ッタ土ハ兩側ヘ積ム

The Proper Order for Building Fortifications

1. Roping Off The Area (Measuring)

2. Digging

3. The soil you dig out should be piled towards the front.

4. Gather sod from an area some distance away.

5. Cover the mounded earth with sod to camouflage the area.

交通（其二）
橋ヤ山地ノ出入口ニハ部隊ガ待機スル場所ヲ作ル
本道
迂回路
掩體

Transportation (Part 1)

Construct places for units to wait near bridges or the start of mountain roads.

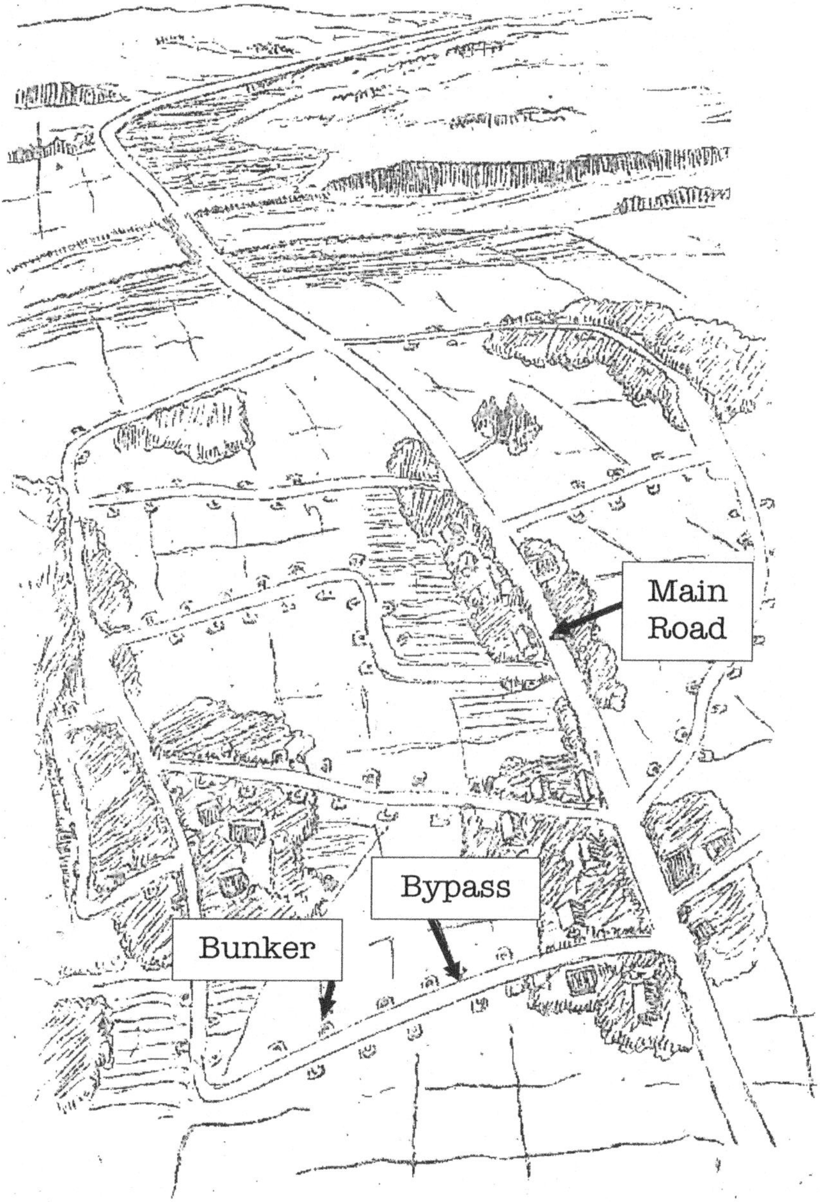

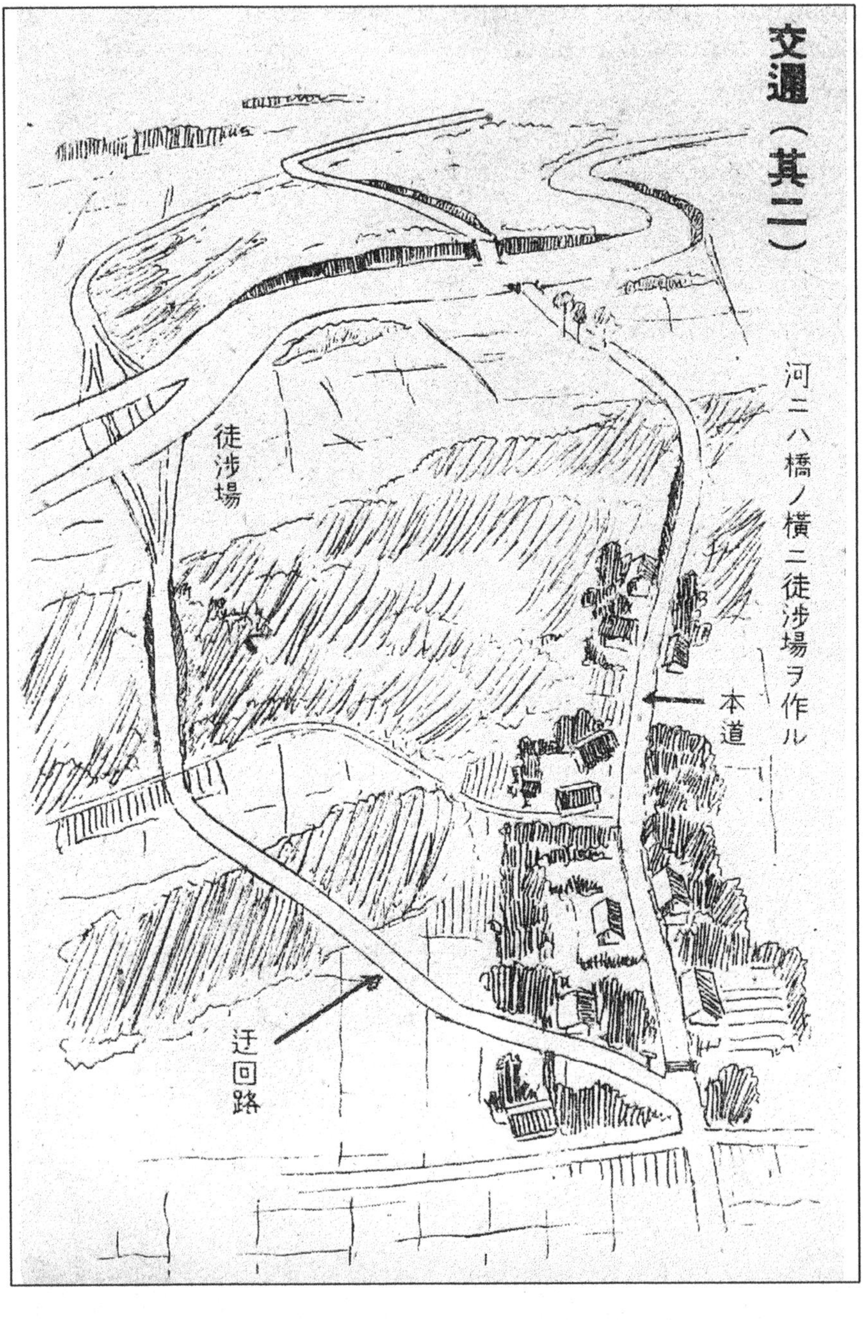
交通（其二）
河ニハ橋ノ横ニ徒渉場ヲ作ル
徒渉場
本道
迂回路

Transportation (Part 1)

Make a place next to a bridge where people can wade across.

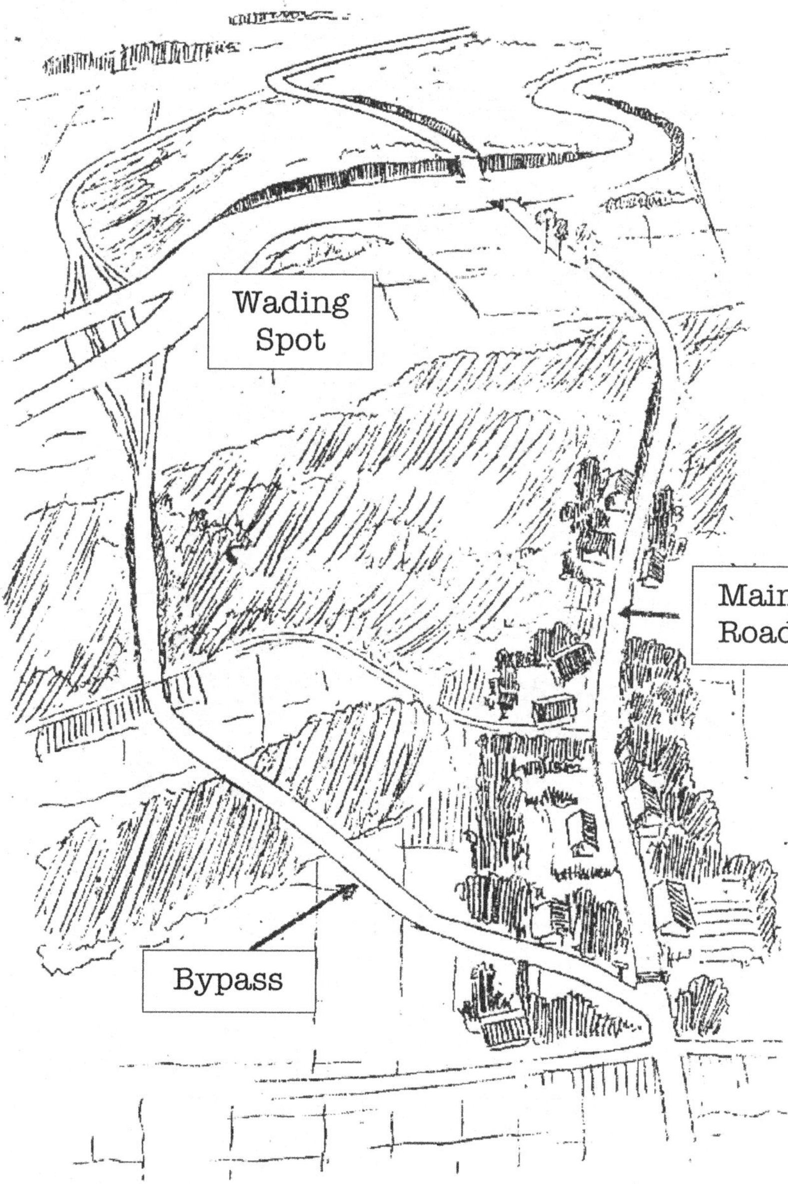

自動車砲車掩體
木ノ下ナド利用シ
偽装ニ氣ヲツケル
0.5
1.0
3.0
2.0
5.0m
水ガタマラヌ
様ニスル
防火用ノ
砂、水、
莚等ヲ準
備スル

Bunker for Automobile or Gun Carriage

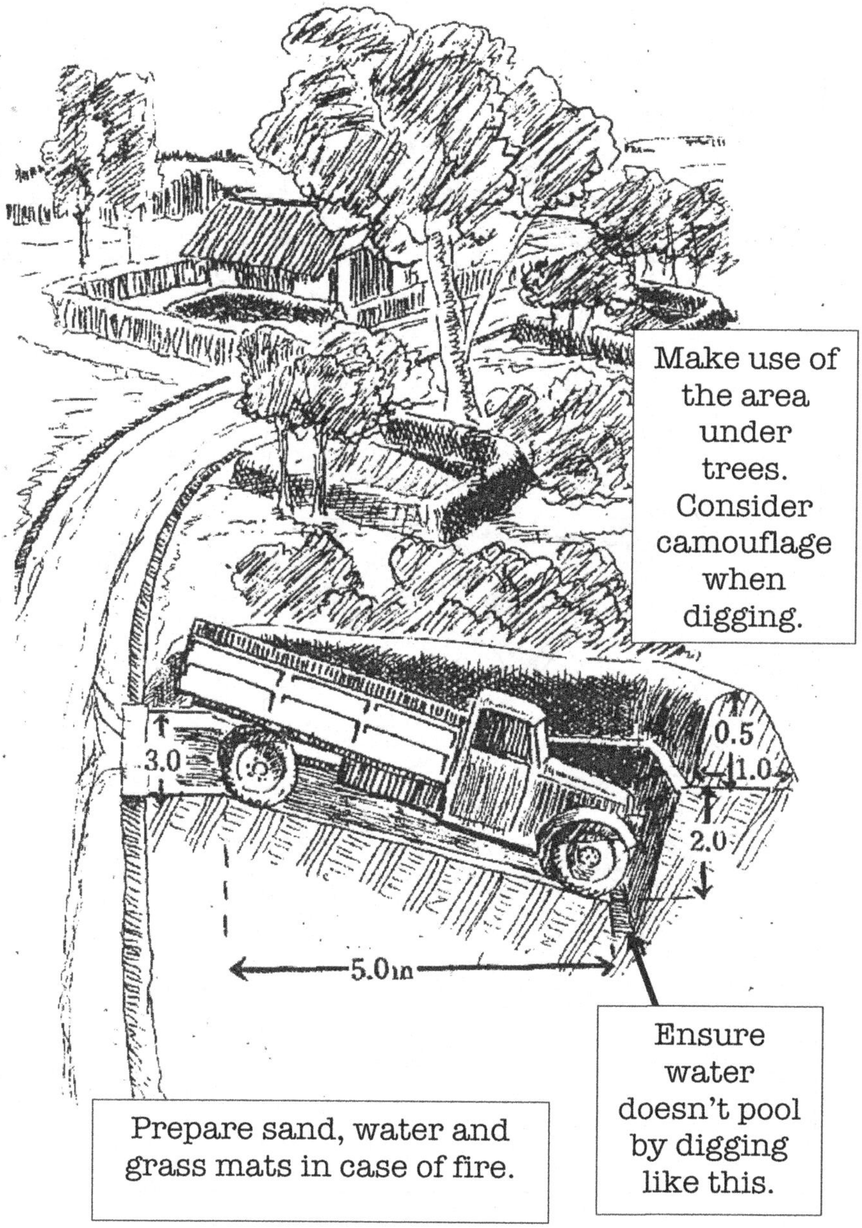

横穴式砲車掩體
3.0
7.0
1.8
自動車ノ場合
3.0
土ヲ捨テル所ニ注意シ
上空カラワカラヌ樣ニスル

Tunnel-Style Gun Carriage Bunker

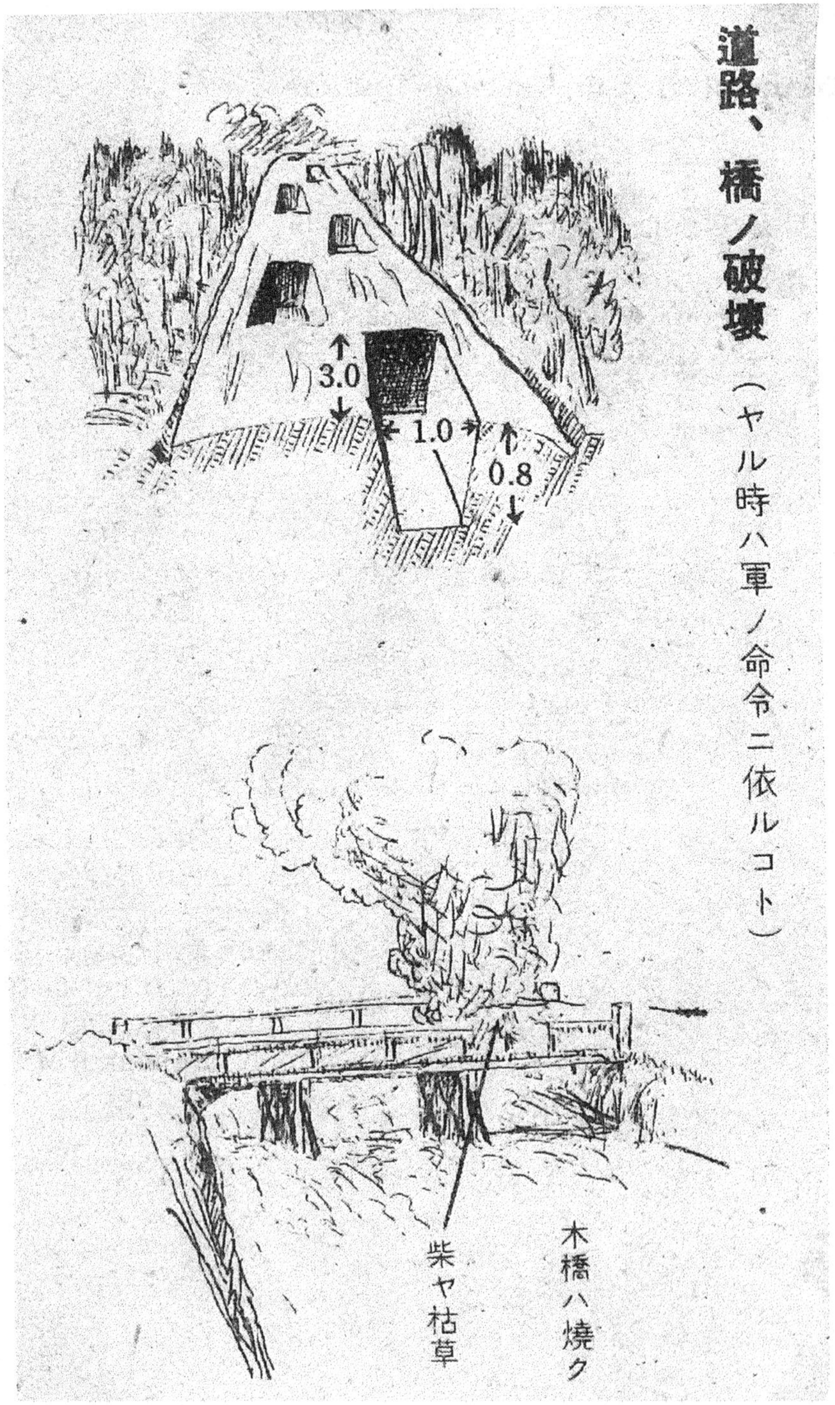
道路、橋ノ破壊
（ヤル時ハ軍ノ命令ニ依ルコト）
3.0
1.0
0.8
木橋ハ焼ク
柴ヤ枯草

Destroying a Bridge or Road

(This should only be done by order of the army.)

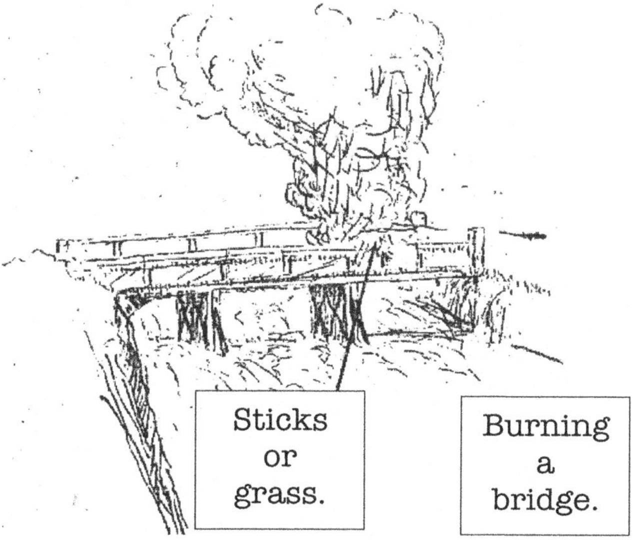

Building Fortifications: The Citizen's Essential Guide

End

Made in the USA
Monee, IL
04 November 2026

67380142R00036